# MINI WEAPONS OF MASS DESTRUCTION®

D0013759

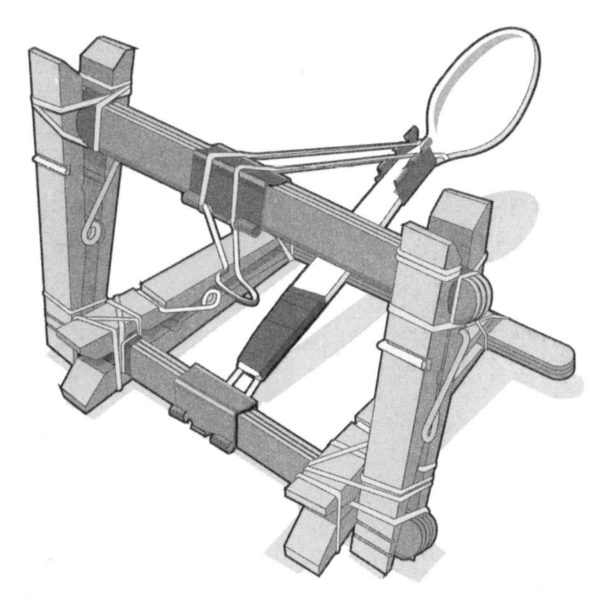

## BUILD IMPLEMENTS OF SPITBALL WARFARE

John Austin

**Library of Congress Cataloging-in-Publication Data**

Austin, John, 1978–

Miniweapons of mass destruction : build implements of spitball warfare / John Austin.

p. cm.

ISBN 978-1-55652-953-5

1. Amusements. 2. Miniature weapons. I. Title.

GV1201.A875 2009

793–dc22

2009026891

Cover and interior design: Jonathan Hahn

Illustrations: Austin Design, Inc.

MiniWeapons of Mass Destruction is a registered trademark of Austin Design, Inc. All rights reserved.

Copyright © 2009 Austin Design

All rights reserved

Published by Chicago Review Press, Incorporated

814 North Franklin Street

Chicago, Illinois 60610

ISBN 978-1-55652-953-5

Printed in the United States of America

20 19 18 17

To all those brave little green army men who experienced horrible melting deaths and are now buried in millions of unmarked sandbox graves. And of course to my father, Steve Austin, who bought me all those courageous PVC soldiers.

# CONTENTS

# INTRODUCTION

*MiniWeapons of Mass Destruction* is a humorous, *MacGyver*-inspired tactical guide that illustrates the full potential of everyday items to be transformed into a menacing miniarsenal.

This bible of forbidden knowledge will prepare you for a zombie uprising or the inevitable alien invasion. To prevent mass hysteria, proper training is essential. Each weapon is cataloged with an easy-to-read bill of materials, step-by-step instructions, and alternate construction methods. And in the final chapter you'll find a small library of simple targets you can use to master your MiniWeapon shooting skills.

This is a book for warriors of all ages. It pushes the laws of physics, inspires creativity, proposes experimentation, and fuels the imagination. Many of the catapults and launchers are great representations of their real-life counterparts, but they cost only pennies, making them great for group exercises and perfect for large-quantity builds.

This book is for entertainment purposes only. Please review the safety page for your personal protection. Build and use these projects at your own risk.

# PLAY IT SAFE

The unexpected can always happen! When building and firing Mini-Weapons, be responsible and take every safety precaution. Switching materials, substituting ammunition, assembling improperly, mishandling, targeting inaccurately, and misfiring can all cause harm. You should always be prepared for the unknown. ***Eye protection is a must*** if you chose to experiment with any of these projects.

Always be aware of your environment, including spectators and flammable materials, and be careful when handling the launchers. Devices that are based on combustion shooters require gaseous fuel; always start with a small amount of the recommended aerosol, then gradually increase the amount to determine what you need for a successful launch. Crossbows and darts have dangerous points, and elastic and latex shooters fire projectiles at unbelievable force and can cause damage. Never point these launchers at people, animals, or anything of value.

It is important to remember that since miniweaponry is homebuilt, it is not always accurate. Basic target blueprints are available at the back of the book; use these to test the accuracy of your MiniWeapons.

This book also features minibombs. Despite their names, they could never be modified to do serious damage. However, they are very loud and can cause hearing impairment or hearing loss. ***Ear protection is recommended*** for all bombs.

Always be responsible when constructing and using miniweaponry. It is important that you understand that the author, the publisher, and the bookseller cannot and will not guarantee your safety. When you try the projects described here, you do so at your own risk. They are *not* toys!

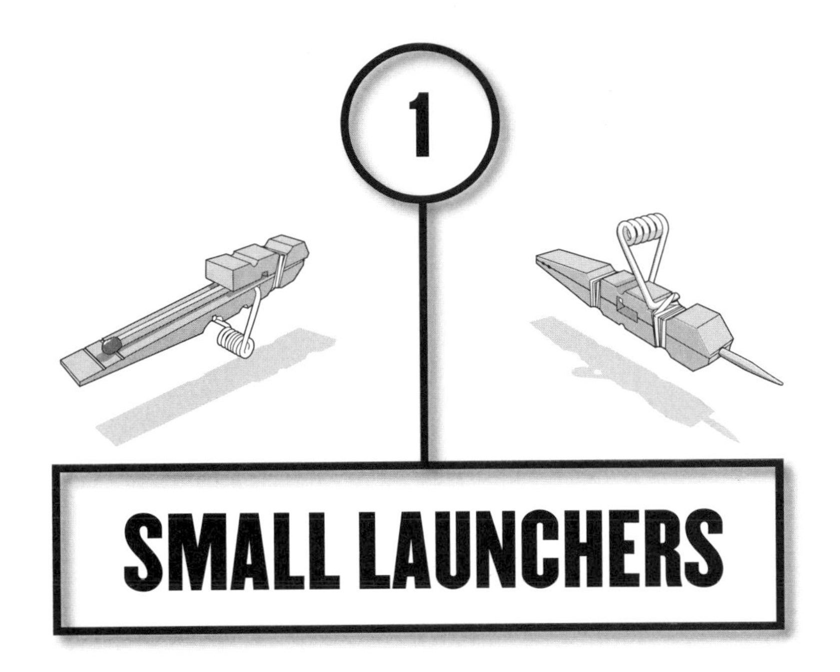

# 1 SMALL LAUNCHERS

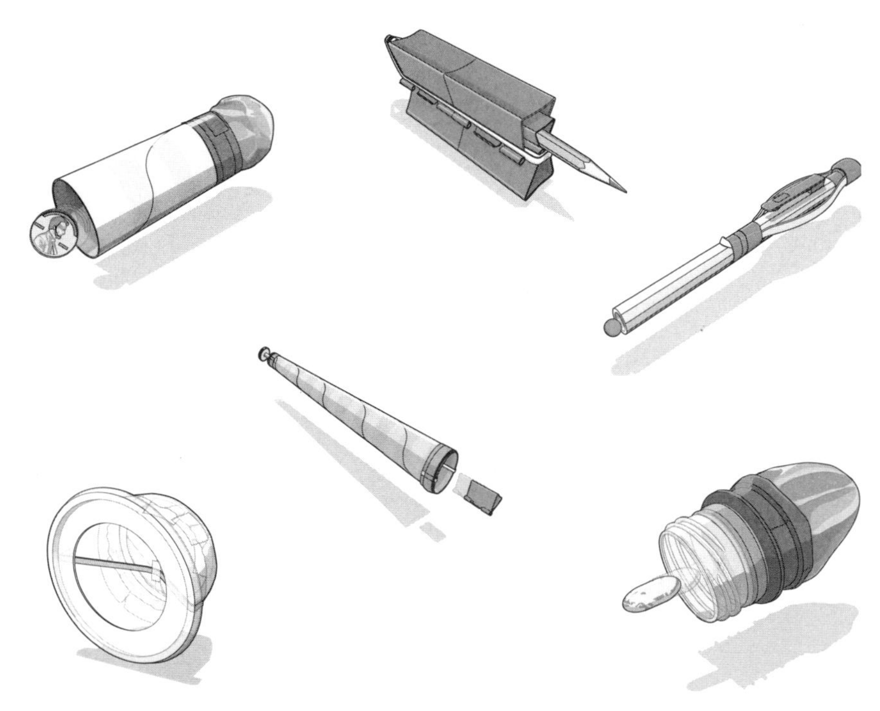

# BB PENCIL

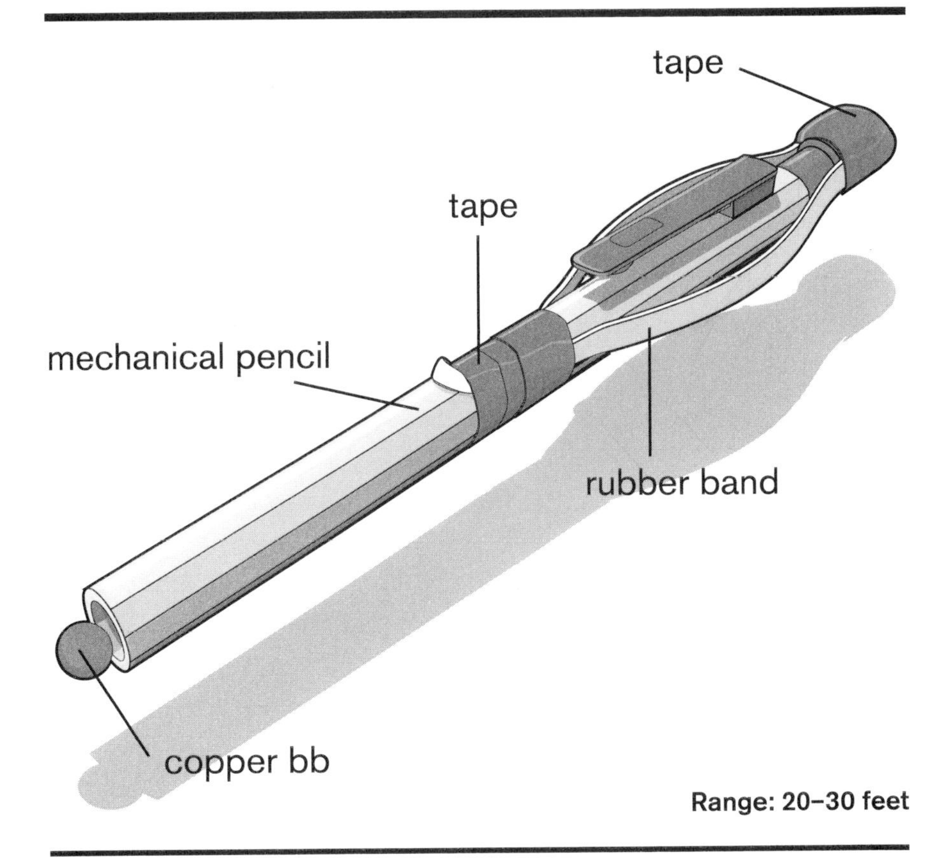


**Range: 20–30 feet**

The BB Pencil is a small, pocket-sized rifle designed to elastically launch a single BB. Requiring limited supplies, this is a great mechanical gun with incredible accuracy. The pencil retains its original form, making it easy to conceal in plain sight—the perfect double-agent tool.

## Supplies

1 inexpensive mechanical pencil
1 wide rubber band
Masking or duct tape

## Ammo

10+ BBs

## Tools

Safety glasses
Scissors (or hobby knife)

# Step 1

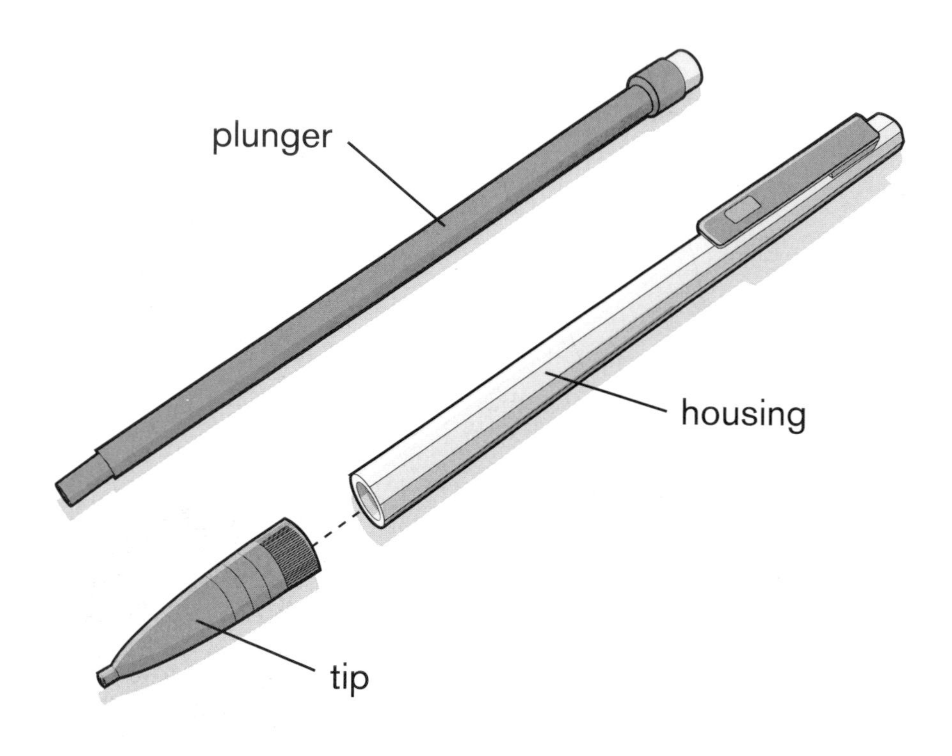


Dissect your inexpensive mechanical pencil using brute strength. Pull out the innards and snap off the pencil tip. If you are unable to perform the decapitation by hand, use pliers. Once removed, the pencil tip can be discarded.

The outer housing of the pencil becomes the barrel of the BB Pencil. Make sure the housing pathway is unobstructed by any plastic fragments that might have broken off during disassembly.

## Step 2

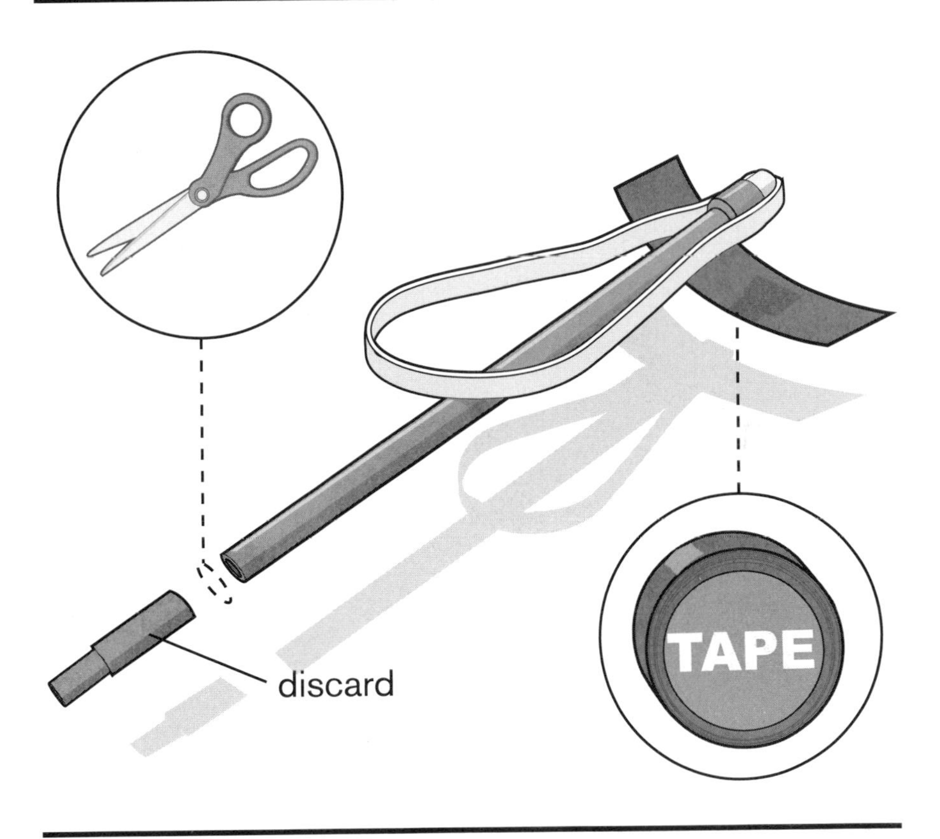


Carefully use scissors or a hobby knife to remove the tapered end of the plunger. This will make room for the BB to sit in the barrel, waiting to be launched, and will prevent it from falling out.

Next, securely fasten a wide rubber band to the eraser end of the plunger with tape. If the eraser is newer, cut a slit in the eraser head to slide the rubber band into for additional support.

## Step 3

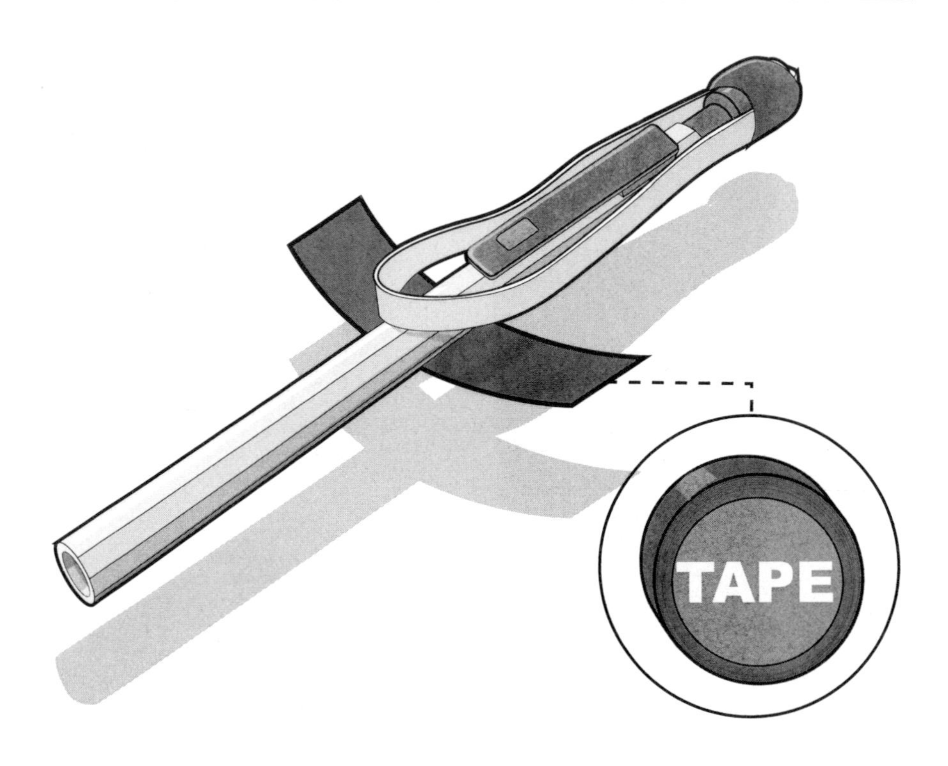


Once you've modified the plunger, slide it back into the pencil housing barrel.

Lay the rubber band on top of the pencil housing. Eliminate any slack in the rubber band and securely fasten it to the pencil housing with tape. Any malfunction during operation will be directly related to the taping of the rubber band, so make sure it's secure.

# Step 4

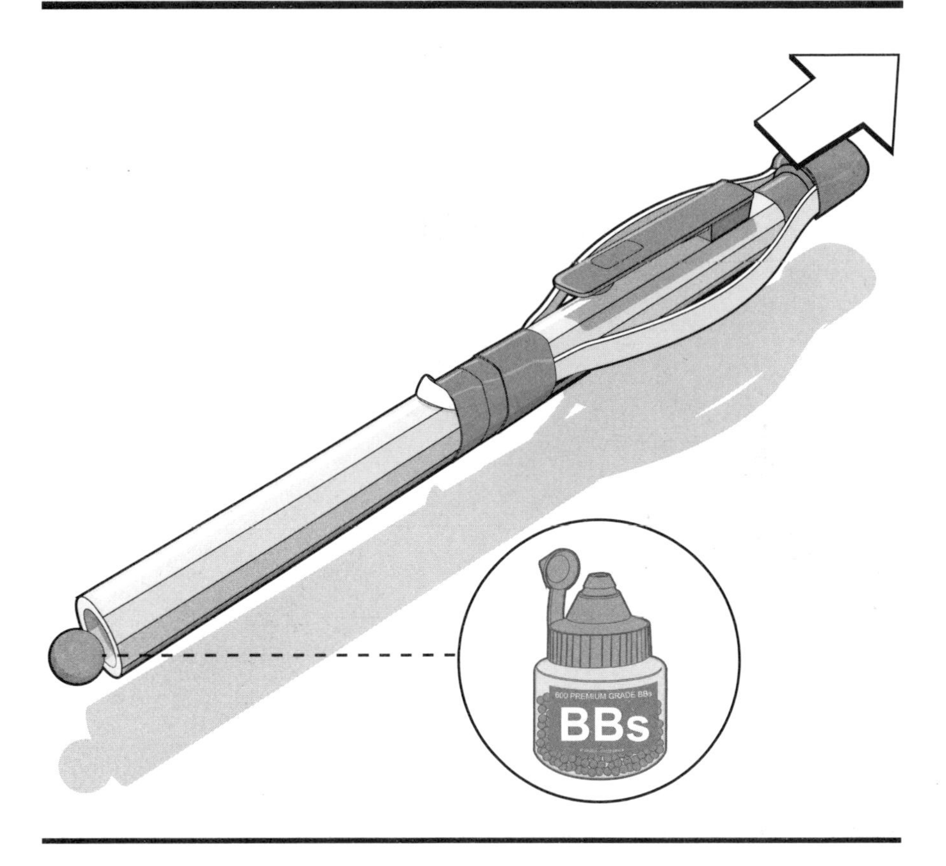


Your BB Pencil is now fully operational. Load a single BB into the muzzle end of the barrel and choose your target carefully. Then pull back the plunger and let it rip!

This mechanical marvel is capable of firing BBs with incredible force and can cause damage. It is important to remember this Mini-Weapon is homebuilt and not always accurate. If you wish to test your BB Pencil, basic targets are available in chapter 7.

## Alternate Construction

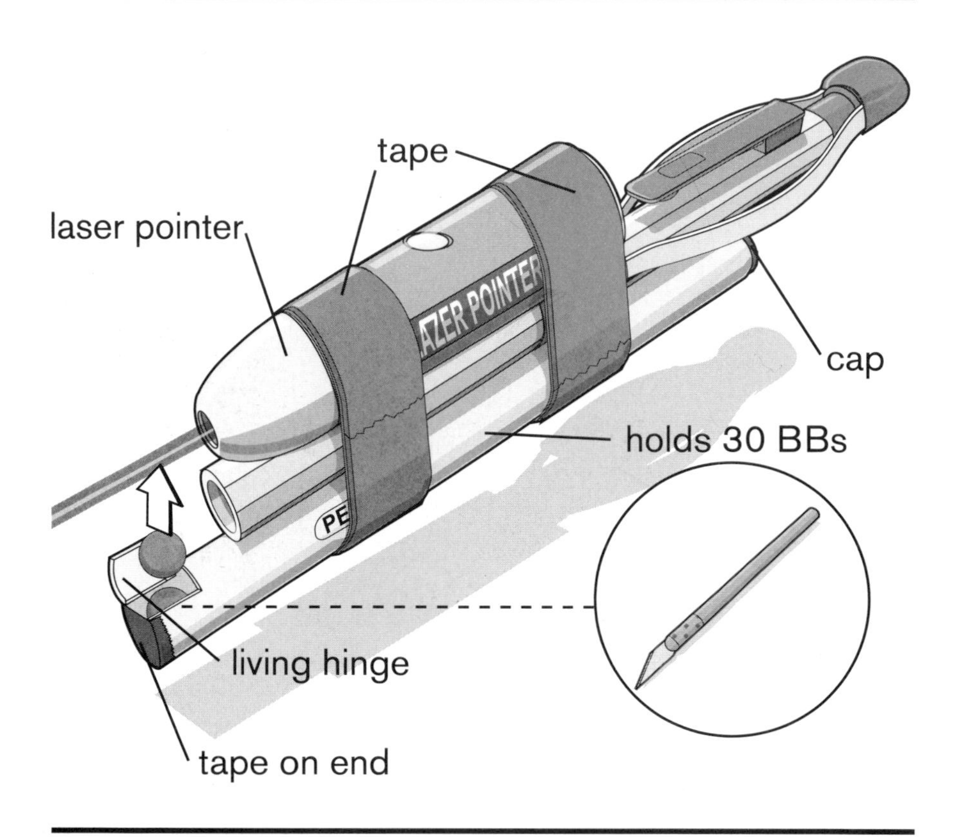


With a few simple additions, you can modify your BB Pencil by incorporating an advanced laser targeting system with additional ammo rounds at your fingertips.

Using an emptied-out pen housing, cut a small BB-sized door at the open end. Then use a small piece of tape to cap the open end. The housing should hold about 30 BBs and can easily be dumped out for a quick reload.

Next, tape both the pen housing and a small, inexpensive laser pointer onto your BB Pencil. Turn on your pointer when you are ready to fire.

# COIN SHOOTER

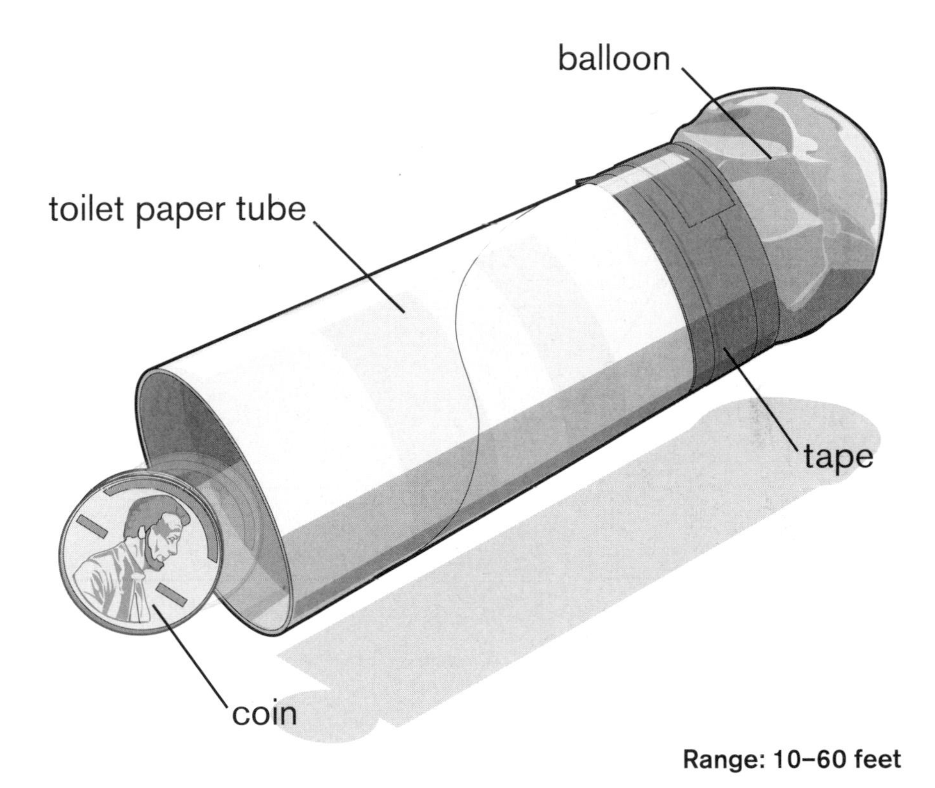


**Range: 10–60 feet**

Everyone on the ground—the party's over! The Coin Shooter is an amazing, inexpensive elastic launcher with an incredible range. Because the gun housing is made from a recycled toilet paper tube and a balloon, you'll not only be firing pennies, you'll also be saving them.

## Supplies

1 balloon
1 toilet paper tube
Duct tape

## Ammo

1+ coins

## Tools

Safety glasses
Scissors

# Step 1

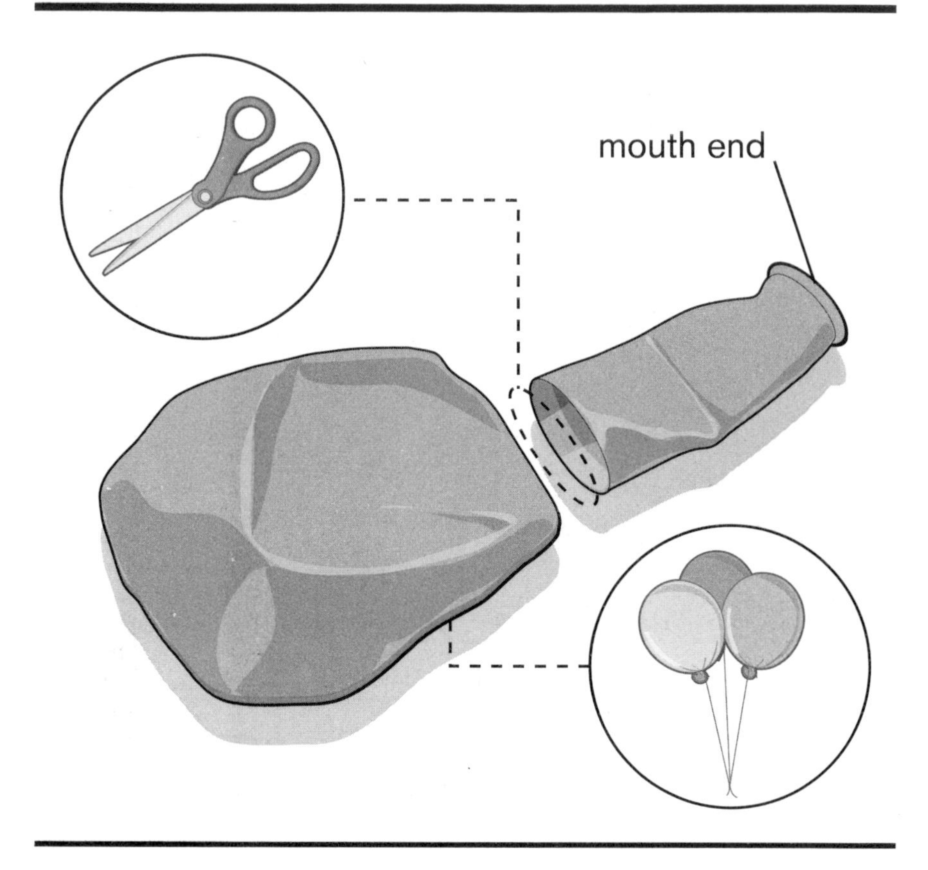


Using scissors, cut one standard-sized latex balloon in half as shown. Throw the mouth end off to one side; you will not be using it for this build.

## Step 2

Now it's time to manufacture the barrel of your elastic cannon. Toilet paper tubes are ideal for this component. If toilet rolls are not available, you can find thick, sturdy tubes inside rolls of paper towels, aluminum foil, and wrapping paper.

Pull the balloon head over the end of the cardboard cylinder. Once in place, tape it securely into position without denting or crushing the tube. Once taped, pull the balloon back to test your adhesive restraint. Add more tape if needed. It is important that you do not alter the tube diameter when securing the tape. This will affect the efficiency of the launcher.

## Step 3

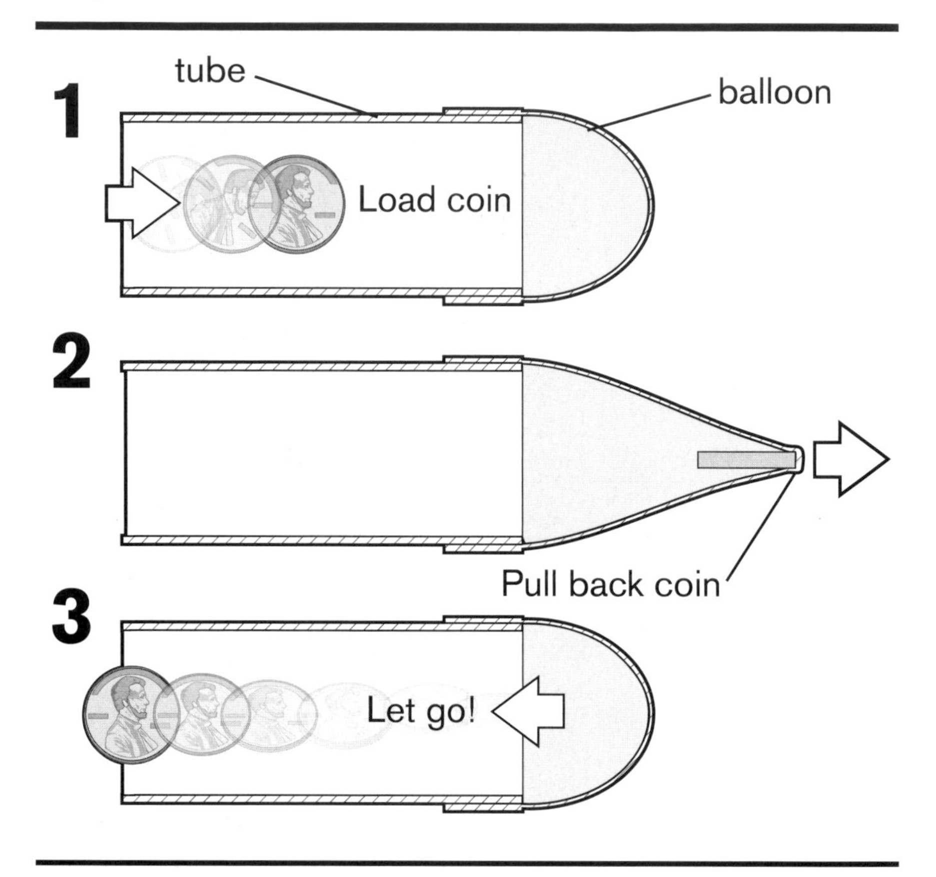


Now load your coin into the muzzle end of the shooter. With your fingers, locate the ammunition in the balloon. Holding the coin between your fingers, pull the balloon back and safely aim the muzzle away from spectators and anything breakable. Release your ammunition from your grip and let it go! The latex will catapult your coin through the barrel at high velocity.

Replacing the short barrel with a longer barrel will help with the accuracy of this firearm. Also, though the gun is designed to launch coins, other low-cost ammunition can be used instead—erasers, marshmallows, spitballs, paper clips, small binder clips, pen caps, peanuts, bouncy balls, and small candies.

Never operate the launcher if the balloon is showing signs of wear.

# BEAN SHOOTER

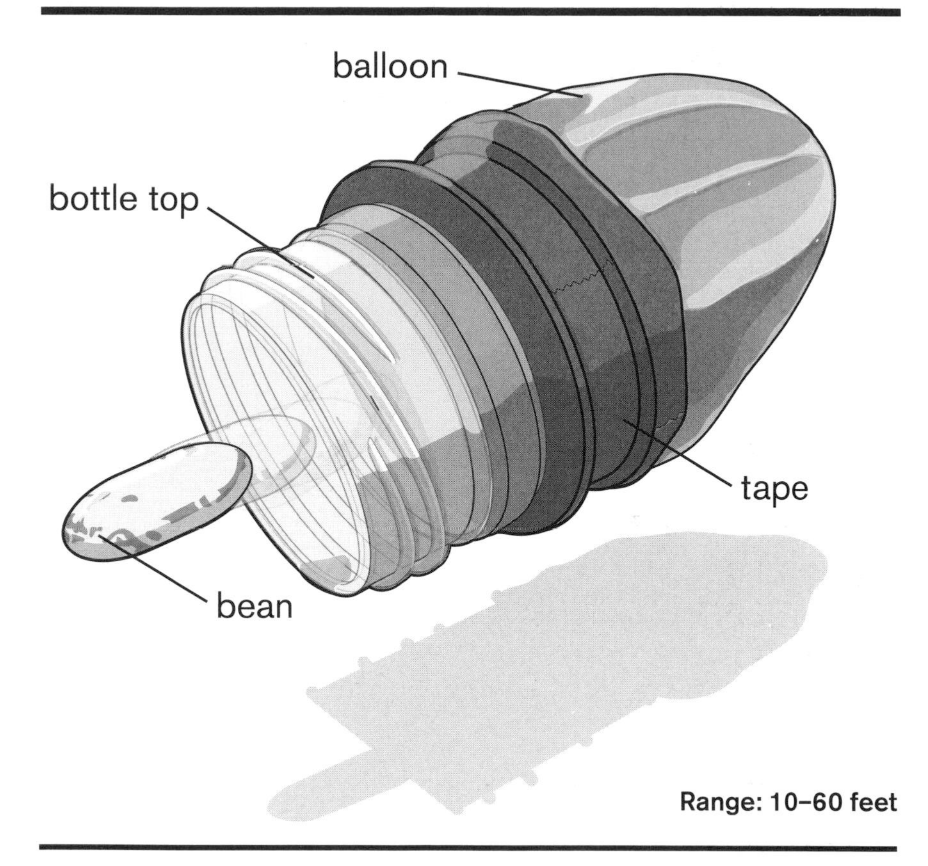


Range: 10–60 feet

The Bean Shooter is the perfect pocket-sized latex launcher. Designed with an unbreakable plastic frame and commanding firepower, it's ideal for the on-the-go hobbyist sharpshooter.

## Supplies

1 plastic bottle
1 balloon
Duct tape

## Ammo

1+ beans

## Tools

Safety glasses
Pocketknife
Scissors

# Step 1

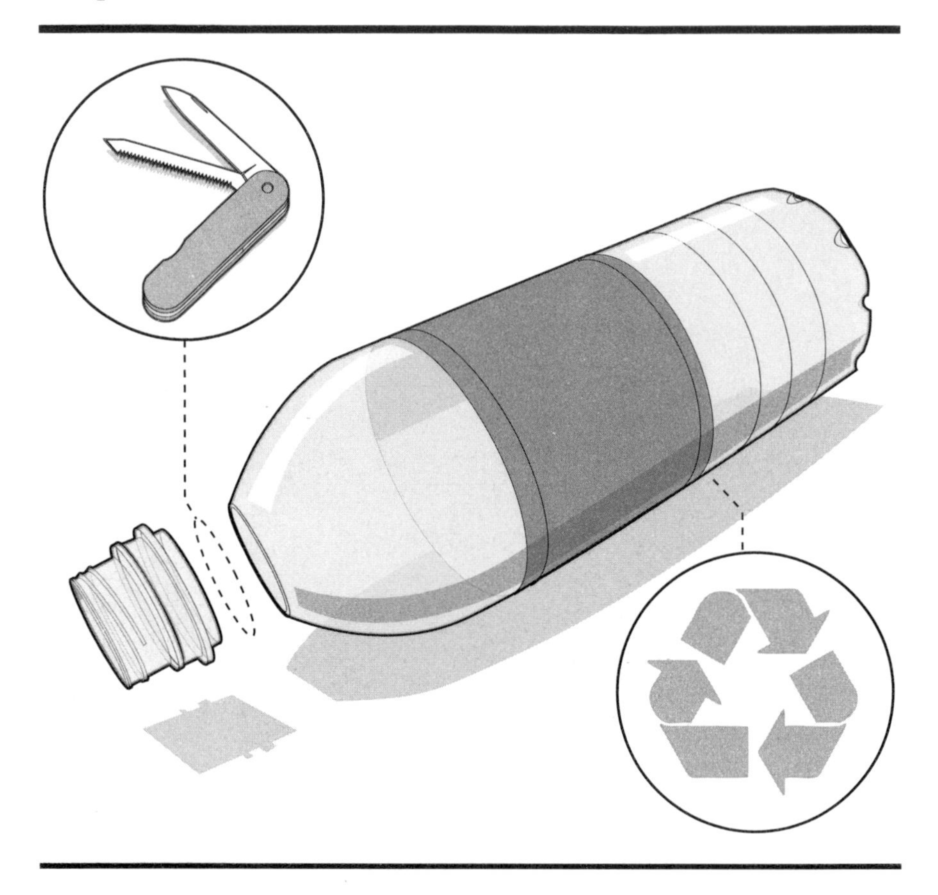

With over 20 billion water bottles sold each year, finding a suitable Bean Shooter frame shouldn't be a problem. How about recycling one of those billons into a musket marvel?

Using one of your various pocketknife blades, saw or cut off the threaded neck of a soft drink bottle. Once you have removed the neck, use your knife to trim off any sharp protrusions that may be left on the cut edge.

## Step 2

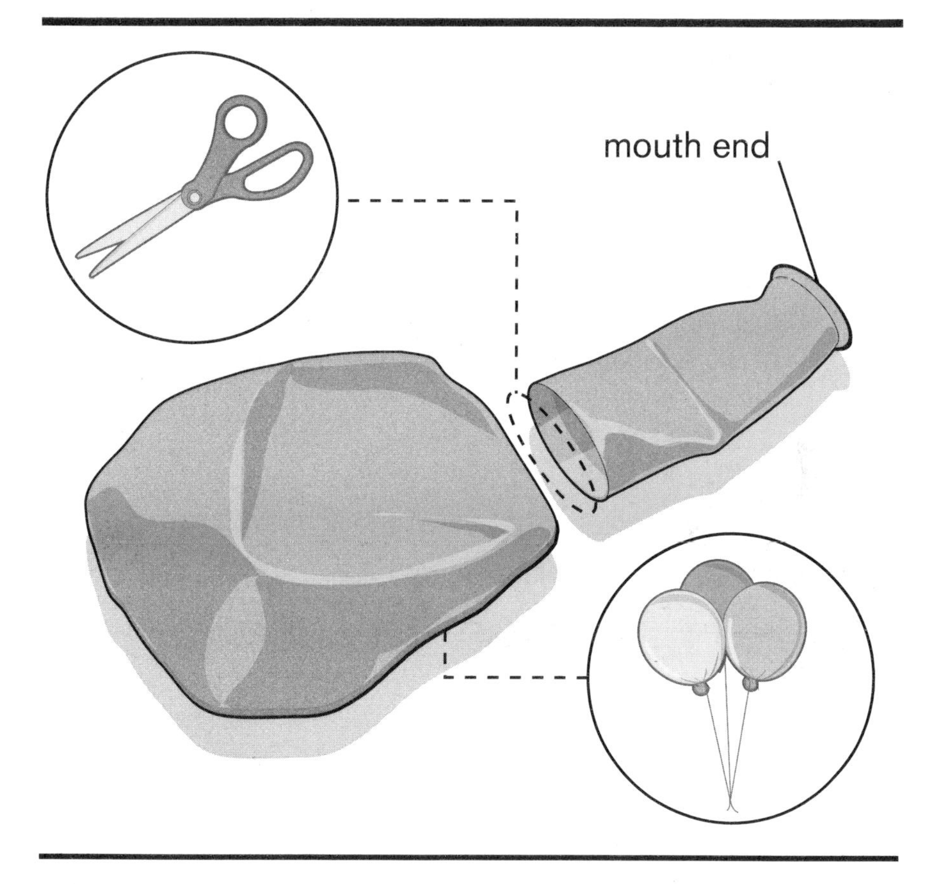


Like the Coin Shooter, the Bean Shooter's power source is a standard-sized latex balloon. Using scissors, cut the balloon in half as shown and discard the mouth end.

Balloons come in many shapes and sizes. A traditional round balloon is recommended for this type of launcher, but feel free to test other balloon shapes for varied results.

# Step 3

Assembly of your Bean Shooter is quick and easy. Take the half section of the balloon and tape it tightly to the bottle top. It is important to be certain that you've cleaned up the cut edge of the bottle in step 1. Sharp edges will cut your balloon and cause a malfunction.

Now, load a bean, eraser, or peanut into the barrel and locate it with your fingers. Once you have a grip on it, pull it back and release. It is important that you pick a safe target to practice your marksmanship.

# CLOTHESPIN SHOOTER

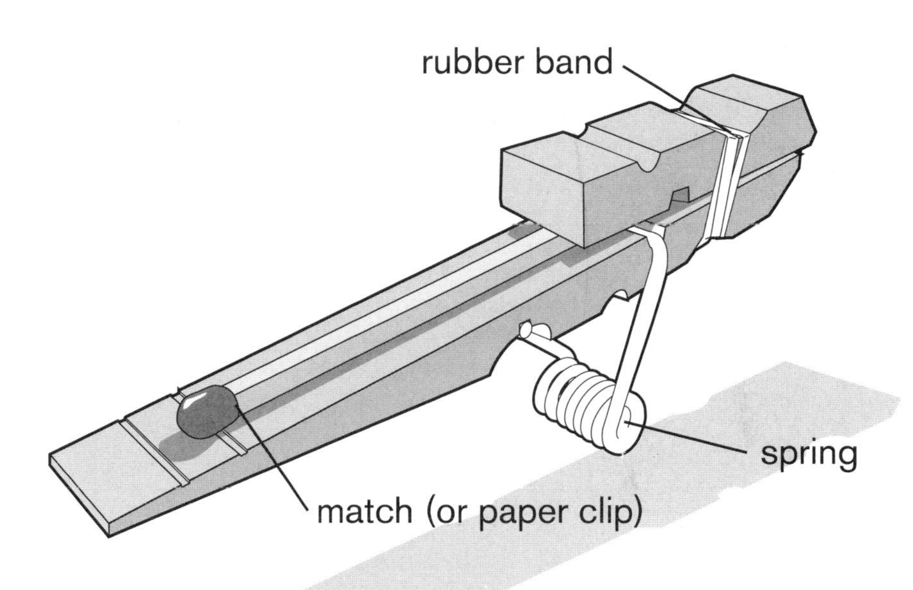


**Range: 10 feet**

Yes, people still use wooden clothespins. You'll be using one to construct a menacing spear gun capable of firing flaming matches. Held like a gun and fired like a gun, this MiniWeapon will have wrongdoers reaching for the sky.

## Supplies

1 wooden clothespin
1 rubber band

## Ammo

1+ wooden matches or paper clips

## Tools

Safety glasses
Pocketknife

# Step 1

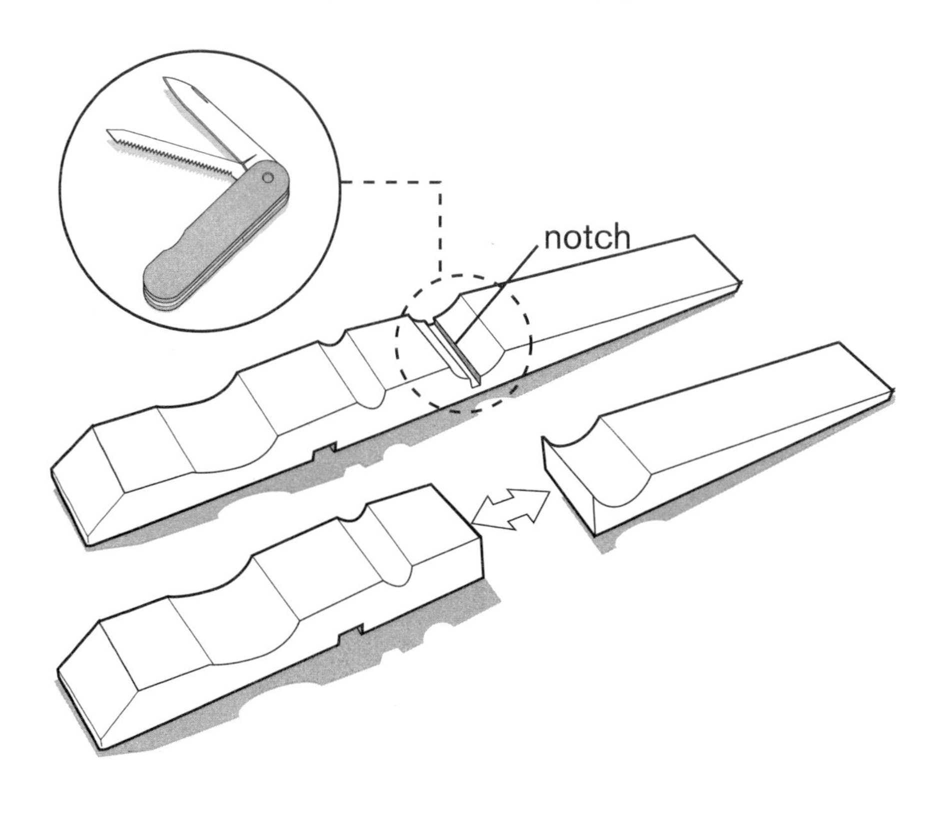


First, disassemble a clothespin and lay the two wooden prongs and metal spring on a cutting surface.

Use one of the blades of a pocketknife to modify the two halves. First, cut a small notch into one of the wooden prongs, as shown. Take your time! Fingers cost more than clothespins. Then, cut the other prong in half at the location shown in the above illustration.

## Step 2

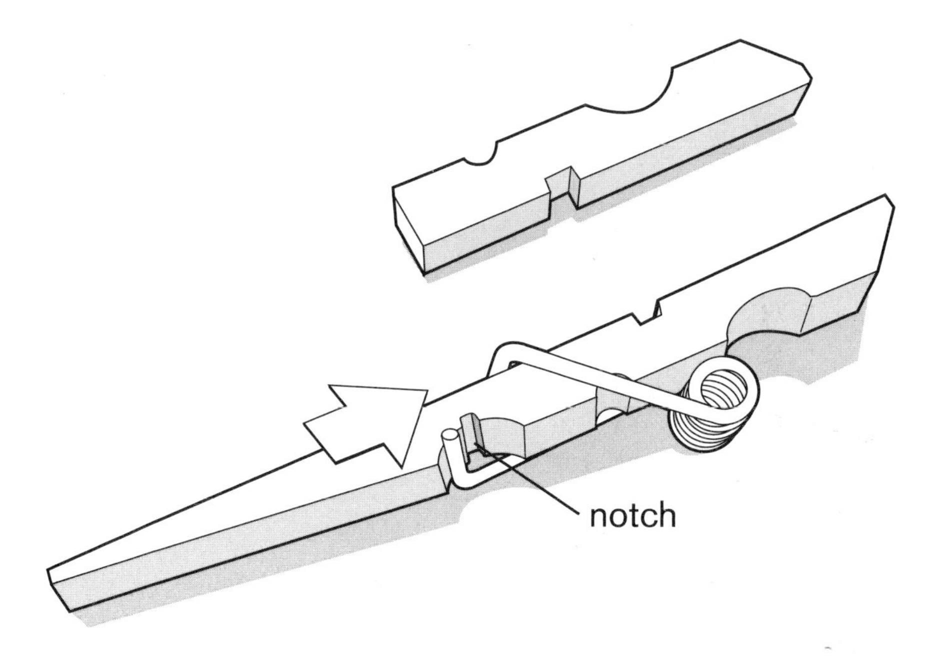


Tightly slide the factory spring back onto the prong with the custom notch you cut out in step 1. Carefully examine the illustration above to make sure the spring is properly oriented. Continue to slide the spring until the bar snaps into the notch. If your notch does not seem deep enough, rework.

## Step 3

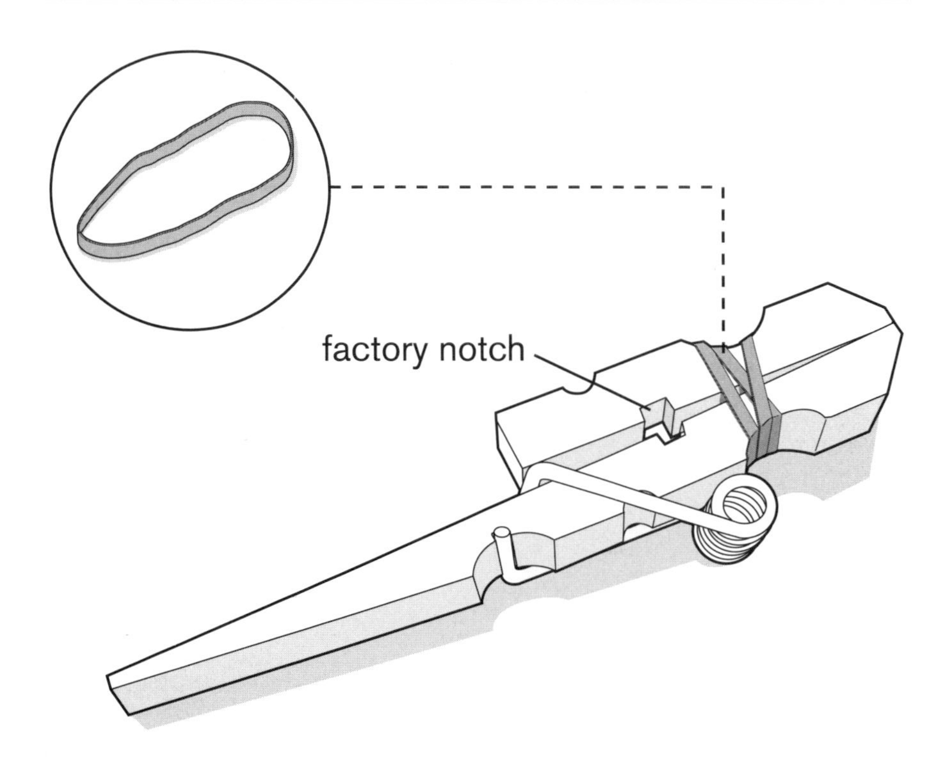


Next, place the shorter wooden prong on top of the longer section. Use the factory notches as a point of reference before sandwiching the two prongs together. Once in place, secure the halves with a rubber band.

Wrap the rubber band just tight enough that you can still move the prongs. To get an idea of how much they should move, keep in mind that the top prong will ultimately slide forward to cock the spring into place.

## Step 4

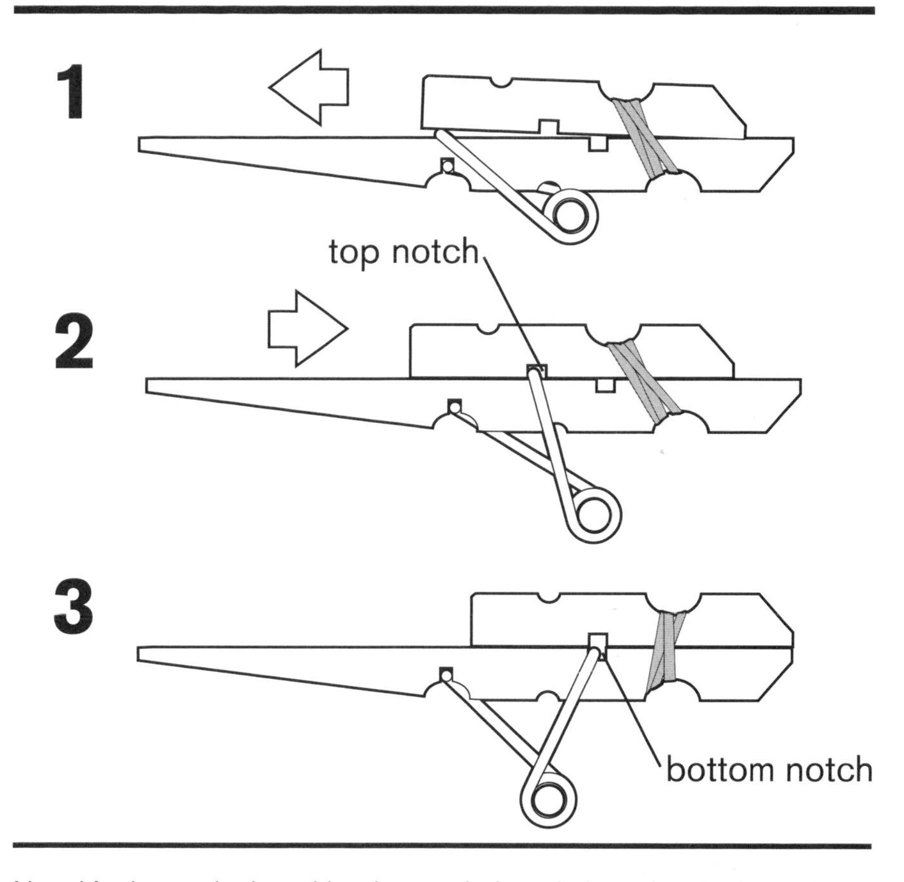


Now it's time to lock and load your clothespin launcher for its first test fire. Push the top prong forward until the spring arm is caught in the top factory notch. Once it is in place, slowly slide the prong back until the spring arm snaps into the bottom notch on the lower prong.

## Step 5

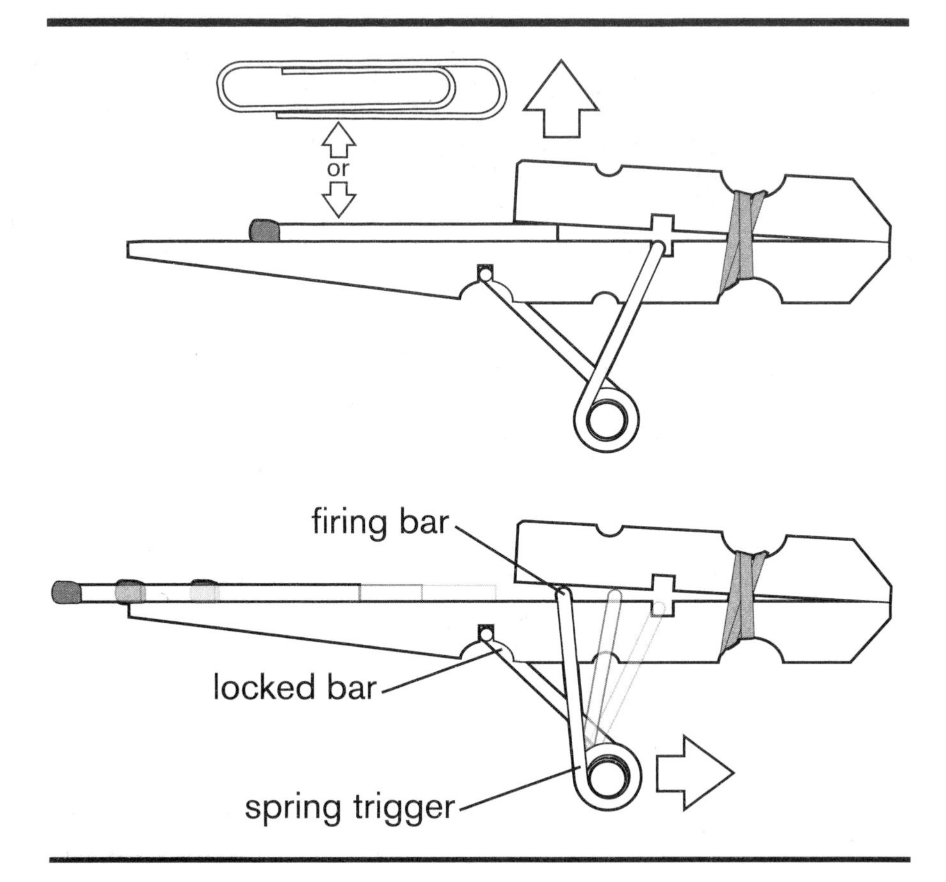


"Reach for the sky, varmint!" Load your chosen projectile into your launcher by lifting the top prong. Wooden matches or standard paper clips are recommended for best results.

Once loaded, pull back the spring trigger to release the firing bar. If the launcher does not fire, but instead the spring arm in the locked bar pops out, rework the locked bar notch to make it deeper. Malfunctions commonly occur because the notch is too shallow.

You can adhere a matchbox striking pad to the prong barrel to light the match when it launches, giving you a flaming arrow. (For a strike-anywhere match, a small piece of sandpaper can be used instead of the striking pad.) For this modification, it's best to reverse the match head. A simpler solution is to light the match before firing. Flaming ammunition is *not* recommended for indoor use. Remove all flammable materials from the area and always wear safety glasses.

# PEG SHOOTER

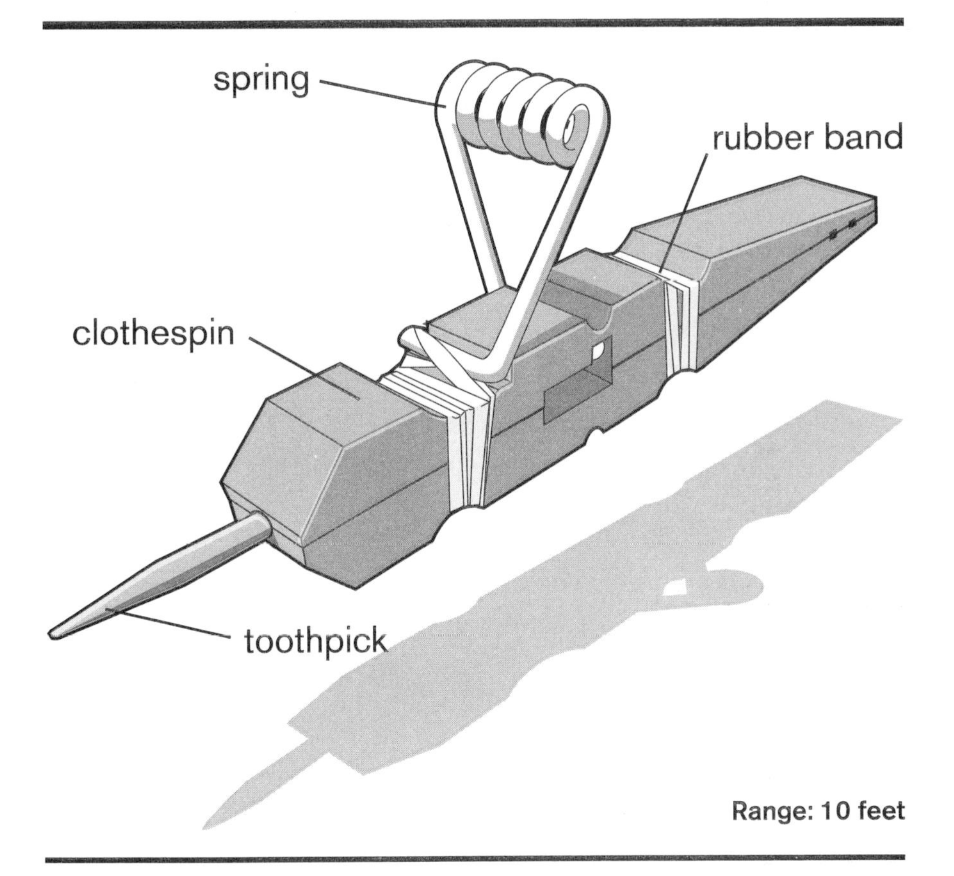


Range: 10 feet

This miniature, spring-powered Peg Shooter is ideal for launching a hailstorm of toothpick firepower. With its palm-sized proportions and quick reload, this shooter is designed for one honorable goal: hanging the laundry out to dry.

## Supplies

1 wooden clothespin
2 rubber bands

## Ammo

1+ toothpicks

## Tools

Safety glasses
Pocketknife

# Step 1

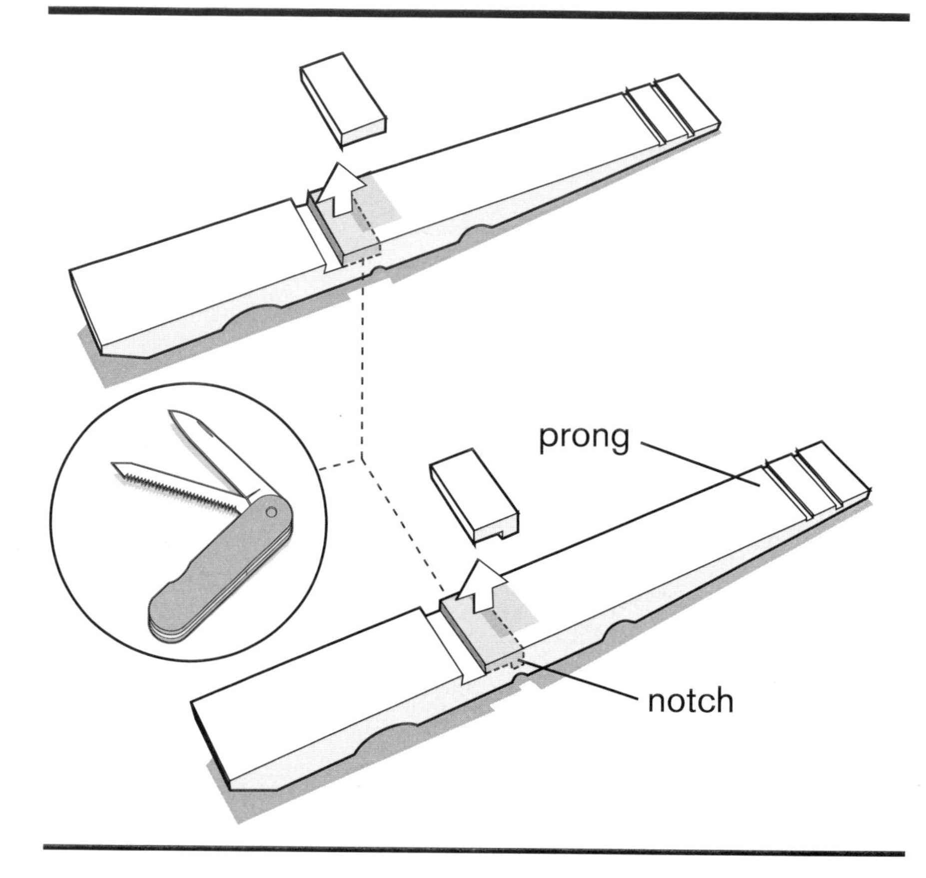


Disassemble a clothespin so that you have two wooden prongs and a small metal spring.

You are going to make similar notches in both prongs. However, one notch will be slightly different than the other prong with a larger notched area. Examine the two illustrations carefully to see the difference. Take your time cutting the notches—craftsmanship is key.

## Step 2

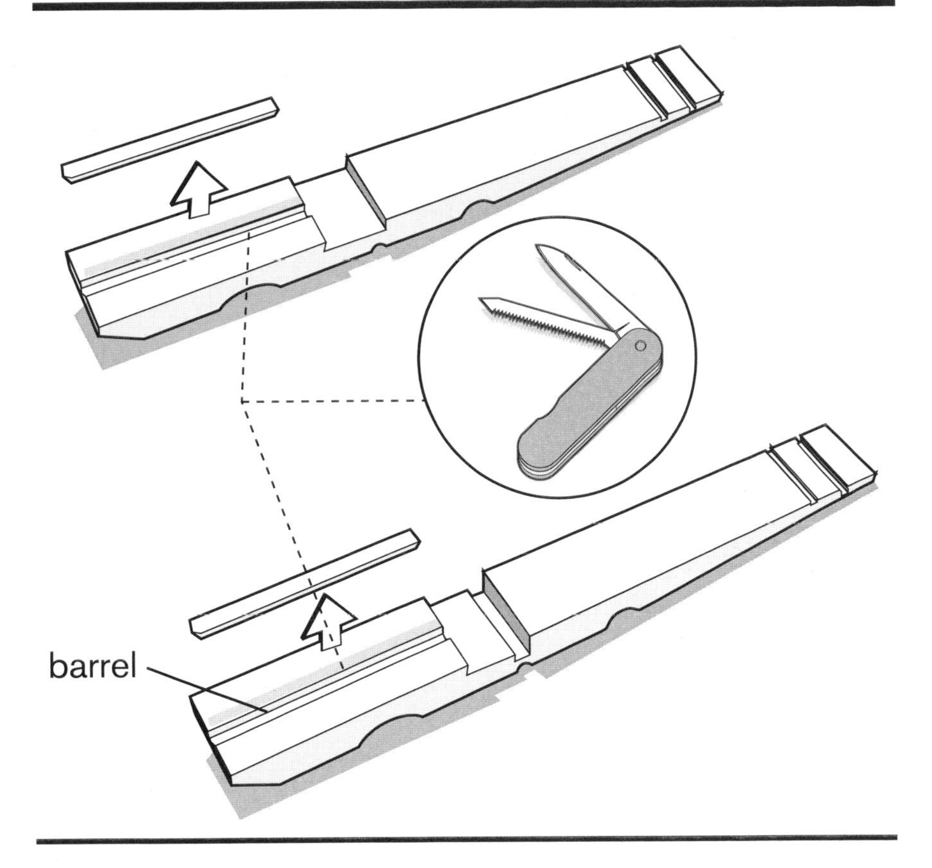


Using a pocketknife, carve a small groove into each prong using a blade that you feel comfortable with. Remember to be careful when handling the blade. Once the prongs are sandwiched together, the grooves should create a channel that is slightly larger than the diameter of a toothpick. This channel will become the Peg Shooter's barrel.

A straight barrel always helps with the accuracy of any gun, so it may help to draw your cut lines on the prongs using a ruler as a guide.

## Step 3

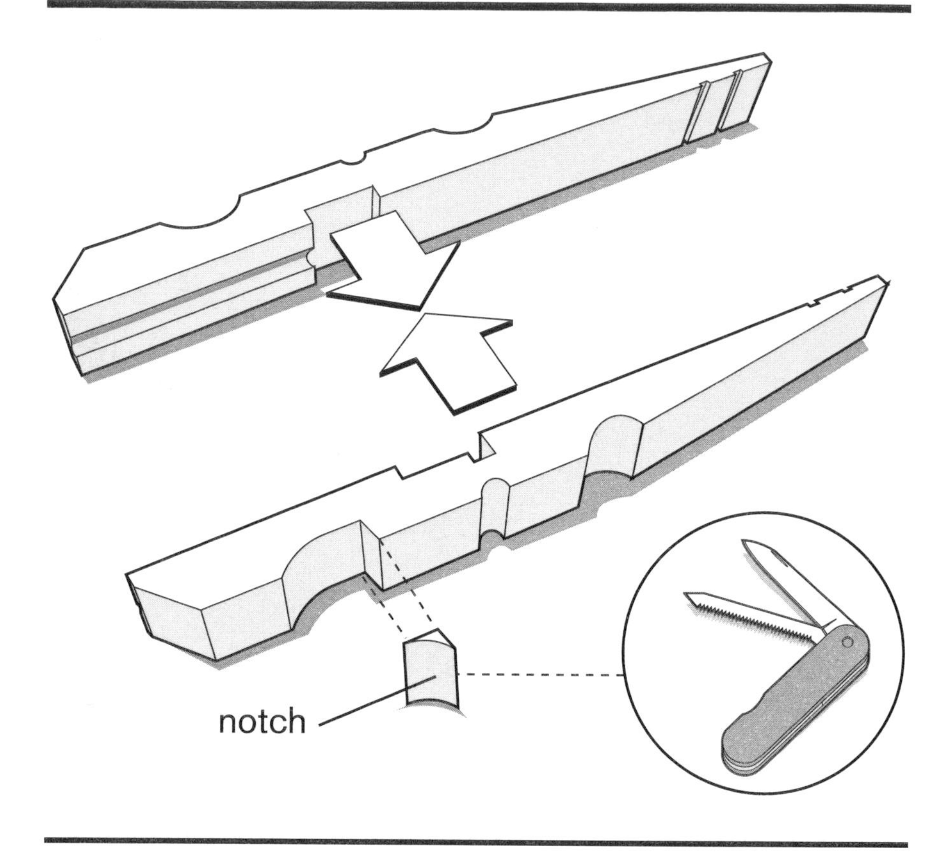


Before sandwiching the prongs together back to back, you will need to remove more material. On the lower prong, at one end of the factory curve, slice the wood to create a 90-degree angle on the curve's back wall, as shown.

## Step 4

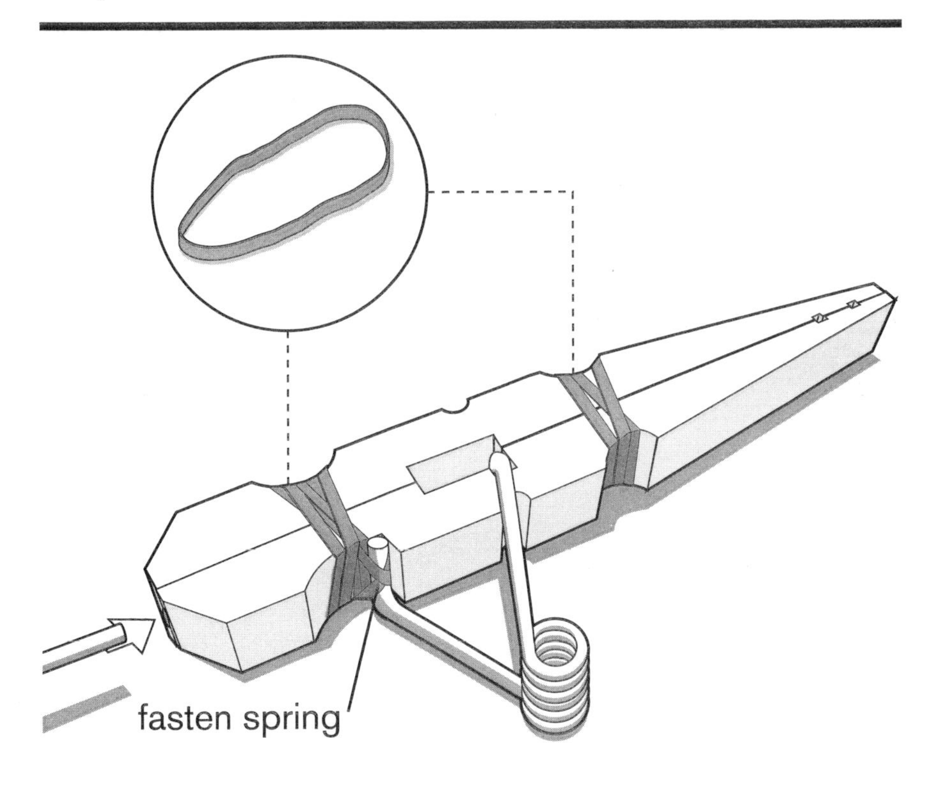


Now it's time to assemble the Peg Shooter. Hold the two prongs with the flat sides pressed together. Once the prongs are aligned, wrap a rubber band around each set of factory curves, at the front and back of the wooden gun. Fine string can also be substituted. Do not obstruct the center of the prongs; this area is reserved for the spring assembly.

Next, wedge the clothespin spring into the wood assembly as shown. One spring arm should be wedged beneath the front rubber band support. (Fastening the front spring under the rubber band will help manage its position.) The other spring arm should be inserted in the middle notch.

## Step 5

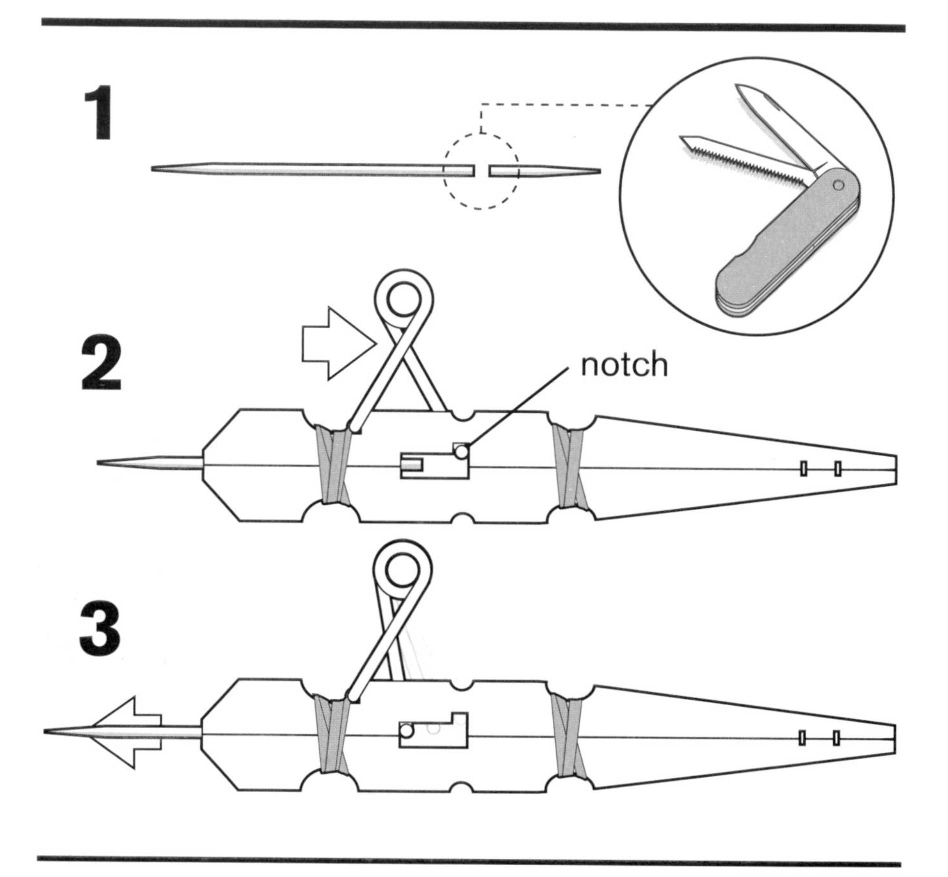


Before the fun and games, you must first prepare the arrow. Take a standard round toothpick and cut off one point (illustration #1). This modified toothpick is your ammunition.

It's now time to lock and load your Peg Shooter. Cock the spring back into its custom notch. Cocking the spring may be difficult and will require one hand to hold the peg gun while the other hand negotiates the stubborn spring into its locked position. Slide the blunt end of the modified toothpick into the muzzle end of the barrel. Safely point your Peg Shooter at a target and, using your finger, pull back the spring so it releases and propels the toothpick forward. If the toothpick does not fire, check your spring to see if your notch is too big. If this is the case, you will have to manufacture a new modified wooden prong with a slightly smaller notch.

Please remember: safety first. You are shooting a fine-tipped arrow out of a homemade launcher. Always wear safety glasses and never point the Peg Shooter at anyone.

# TUBE LAUNCHER

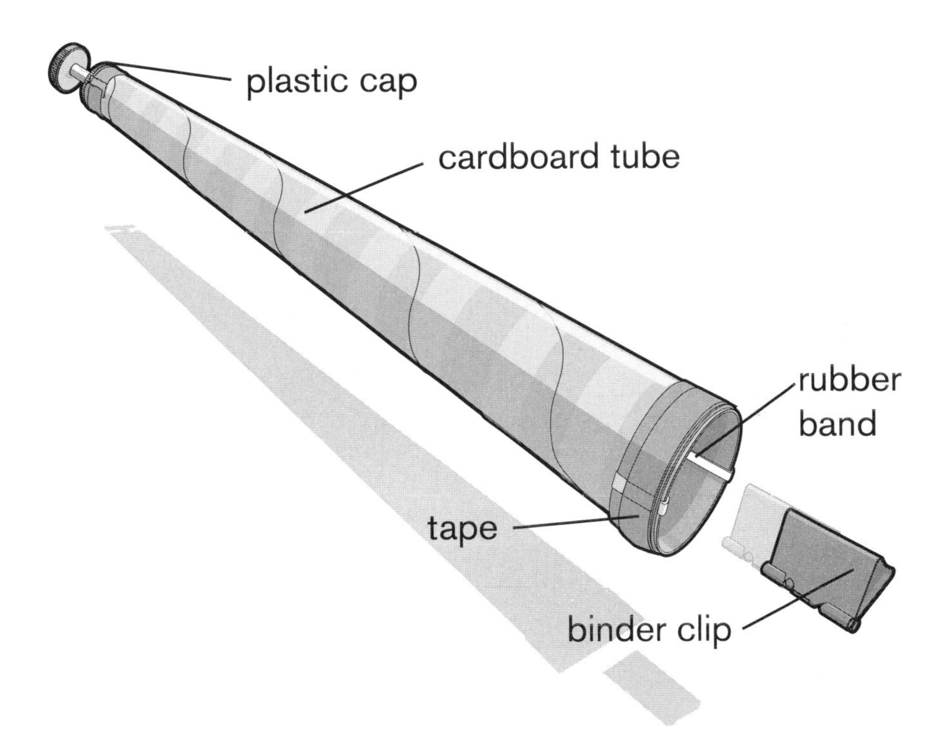


**Range: 20 feet**

The Tube Launcher is also known as the Rock Launcher. Its universal design makes it a great launcher for almost any ammunition.

## Supplies

1 cork
1 paper clip
Duct tape
2 wide rubber bands
Cardboard tube
String
2 plastic caps (from milk cartons)

## Tools

Safety glasses
Scissors
Pocketknife

## Ammo

1+ small binder clips (19 mm)

## Step 1

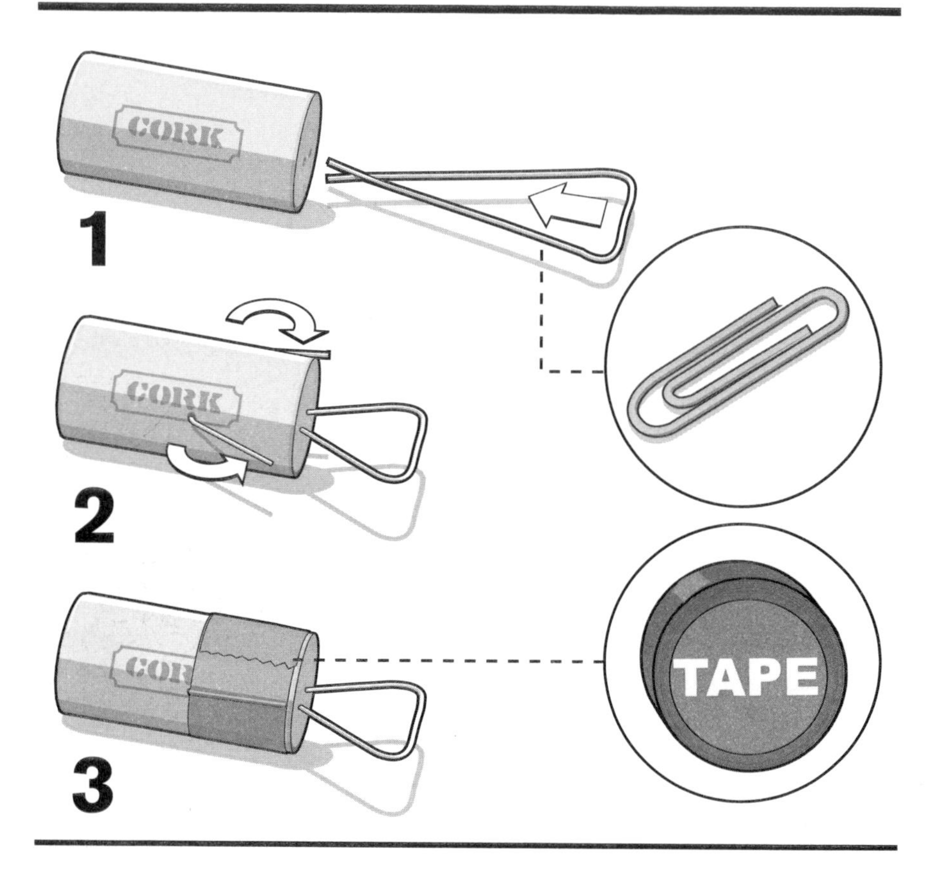


First, select a cork that fits easily inside your cardboard tube. You want to have some clearance on the walls, because you will be adding additional diameter to the cork.

Straighten a paper clip and then bend it into a U shape as shown in illustration #1, above. Next, slowly push both ends of the paper clip into the cork. Aim the ends of the paper clip so that they poke out of the sides of the cork around its midpoint. Bend the ends upward (illustration #2, above) to prevent the paper clip from coming out.

Finally, use tape to wrap up the ends of the paper clip so they do not scrape against the inside of the tube. Test your cork by sliding it through the tube before proceeding to the next step. Rework the paper clip's positioning if the cork jams in the tube.

## Step 2

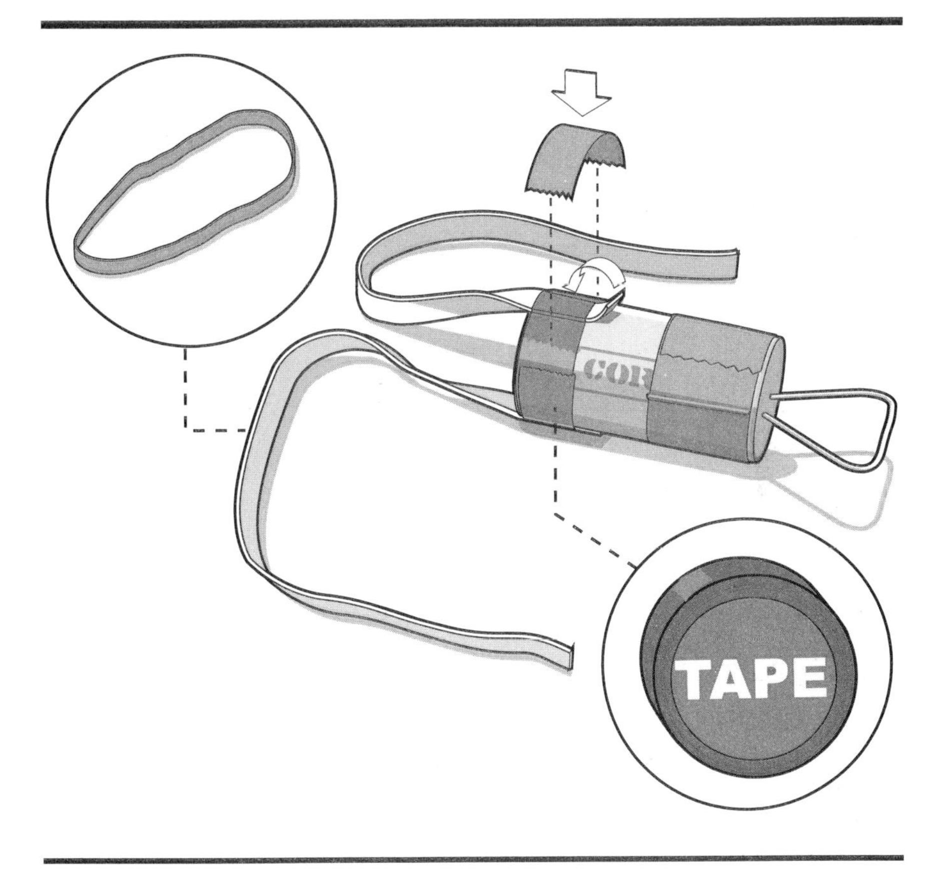


Now it's time to add the elastic firepower to this Tube Launcher. Using scissors, cut both wide rubber bands to create two pieces that are equal in length. Tape the rubber band pieces to the end of the cork opposite the paper clip. When taping the bands it is important to leave their ends sticking out a bit so you can fold them back and tape them a second time. This will help prevent the rubber bands from stretching out of their tape restraints. This is one area that is prone to malfunction, so do a thorough taping job here. For additional holding power, use heavy-duty staples.

## Step 3

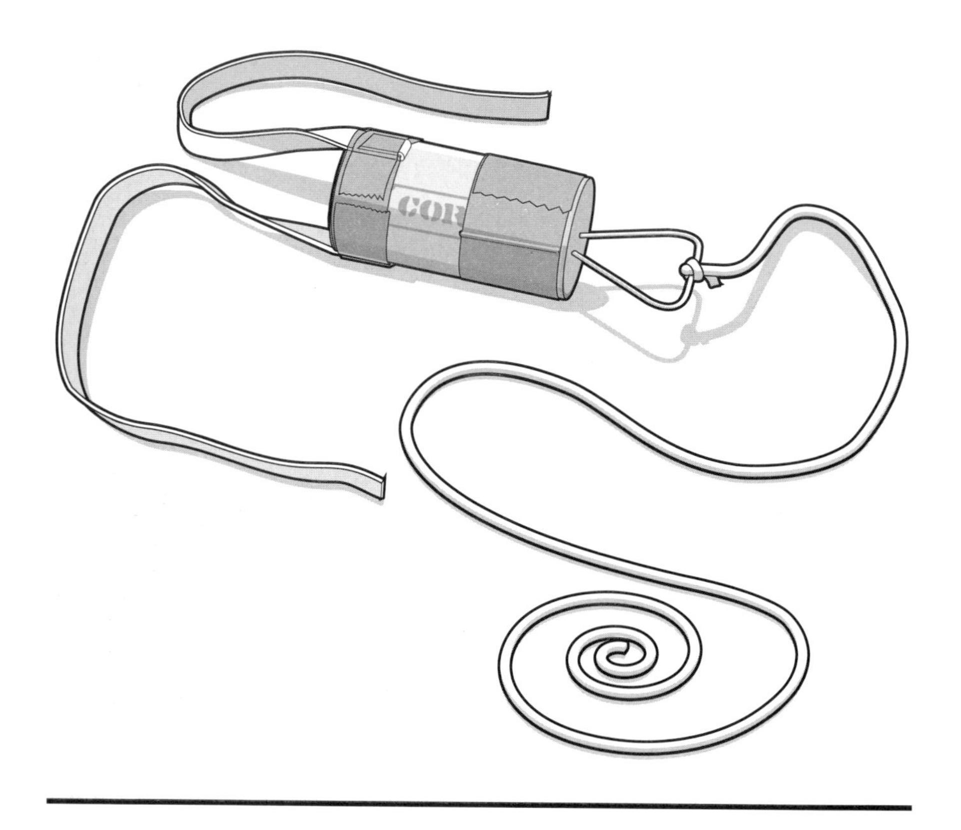


Cut a piece of string that is longer than the length of the cardboard tube. Tie one end of the string to the paper clip. This string will be subjected to tremendous force when you operate the Tube Launcher, so it is important that it has some pulling strength. Double-check your knot. You don't want it to give way while the cork is in the tube. Disassembling the launcher at that point would be a pain . . . so *triple* check it.

## Step 4

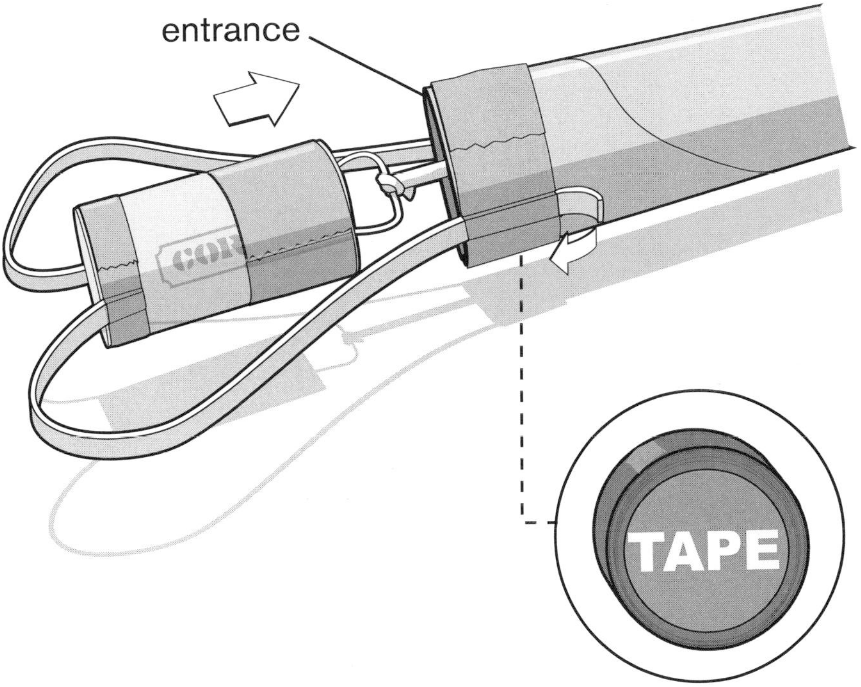

entrance

TAPE

Now it's time to install the components into the cardboard tube. Slide the string and cork assembly into the tube—string and paper clip first. Tape both ends of the rubber bands parallel to each other on the outside of the tube, as shown. Then fold the rubber bands back toward themselves and put additional tape over them. These areas also experience tremendous pressure, so it is important to tape them well to avoid malfunction.

## Step 5

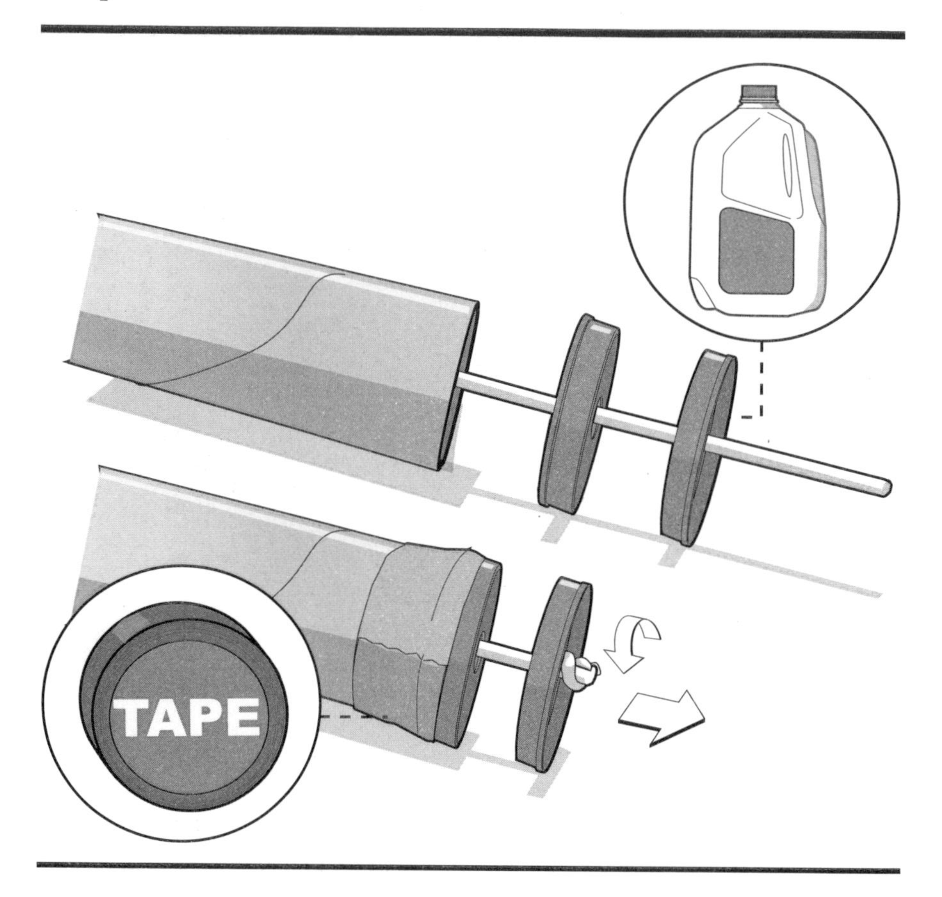


Now that the entrance assembly is finished, it's time to work on the other end.

First, locate two large plastic milk jug caps. Using a pocketknife, cut a small hole in the center of each. Mirroring one another, back-to-back, run the string through the center of the caps. Fasten the cap closest to the tube with tape.

Now, slowly pull the string until you feel some tension from the rubber bands. Tie a few knots at the end so the string can't pass through the hole in the cap. Depending on the hole size, you may want to add a large paper clip to the end of the string as well.

Load your ammunition and pull back on the outermost jug cap. Release the cap and watch your ammo fly!

# MAUL GUN

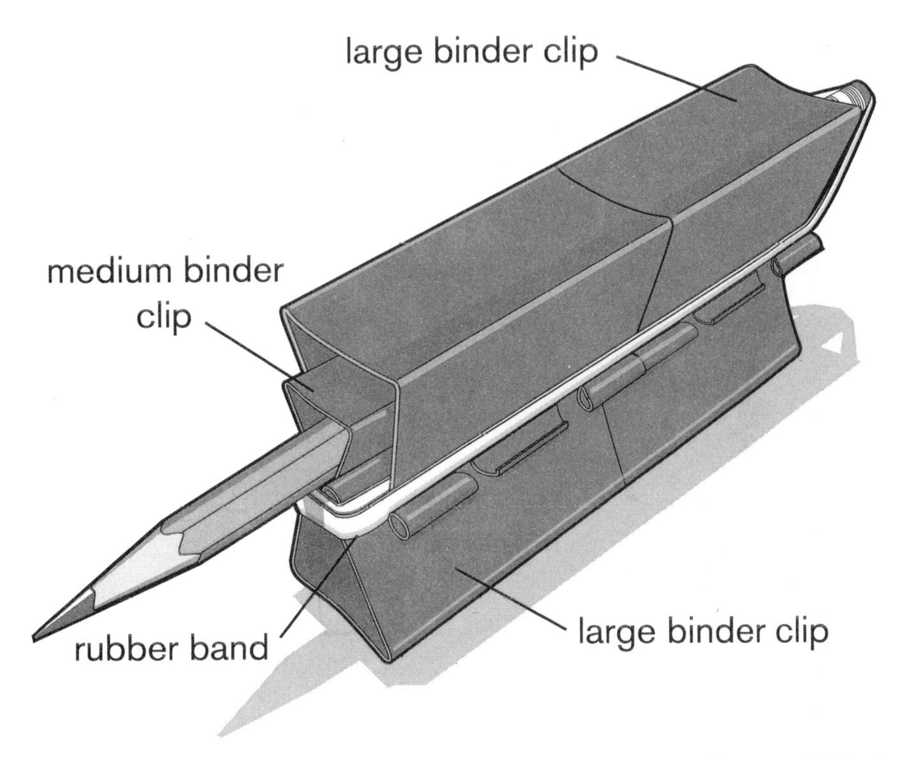


**Range: 20 feet**

The Maul Gun is a feared and dominating piece of hardware. Its raw power can pierce a defenseless aluminum can with ease. But because it's so powerful, it is also very dangerous and should be operated with care. The Maul Gun is not exceptionally accurate because of the somewhat random directions its pencil arrow exits the barrel. For this reason, the Maul Gun is great for either short-distance accuracy tests or long-distance range contests. Never place your intended target in front of a backdrop made of breakable materials such as glass, thin wood, or ceramic.

### Supplies

3 medium binder clips (32 mm)
4 large binder clips (51 mm)
2 wide rubber bands

### Tools

Safety glasses

### Ammo

1 pencil

## Step 1

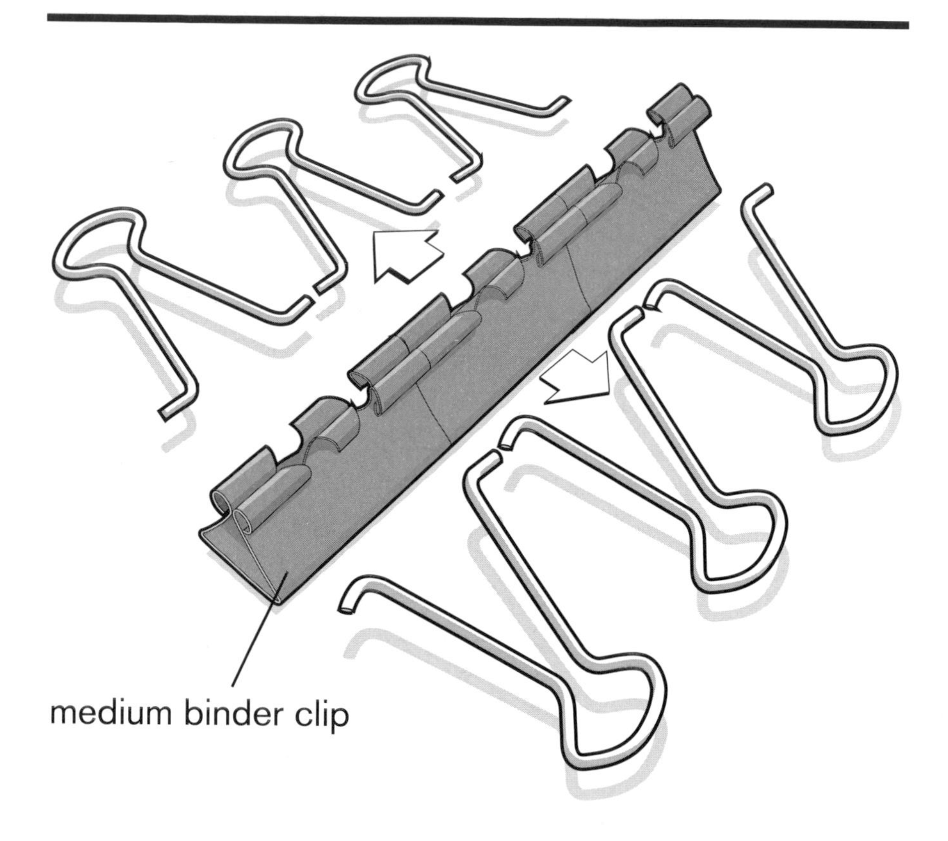


Remove the metal handles from three medium binder clips (32 mm) by squeezing the handles from the sides, then pulling them out. These metal handles can be discarded or recycled, for they are not used in the Maul Gun. Arrange the three binder clips in a line, facing upward.

## Step 2

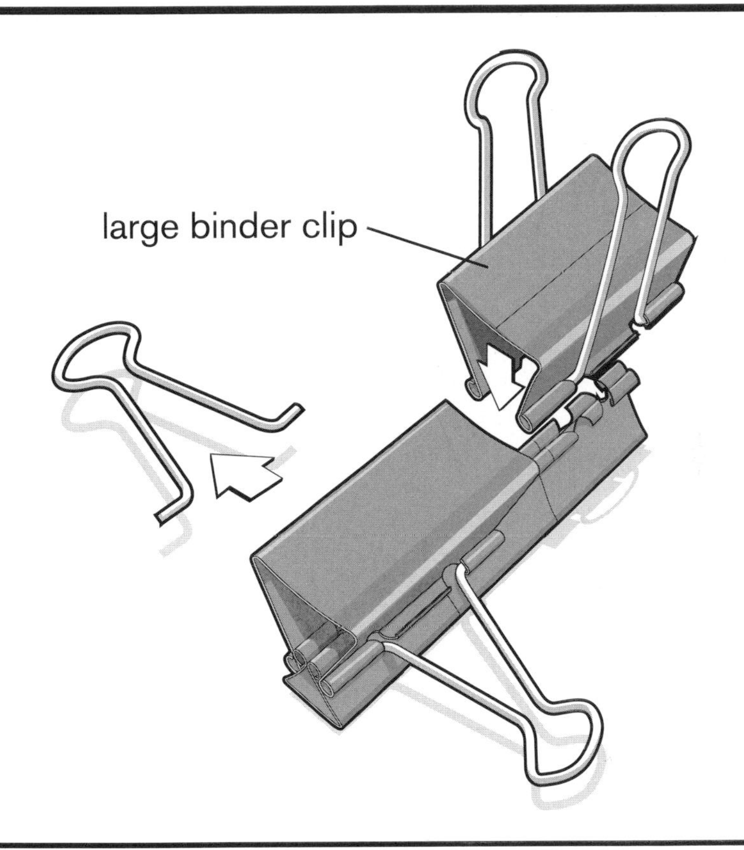


Next, clip two large binder clips (51 mm) onto the row of medium binder clips. Once hooked together, the assembly should be completely attached so you can easily lift it as one piece.

Remove the metal handles from the large binder clips only after you've clipped them securely in place.

## Step 3

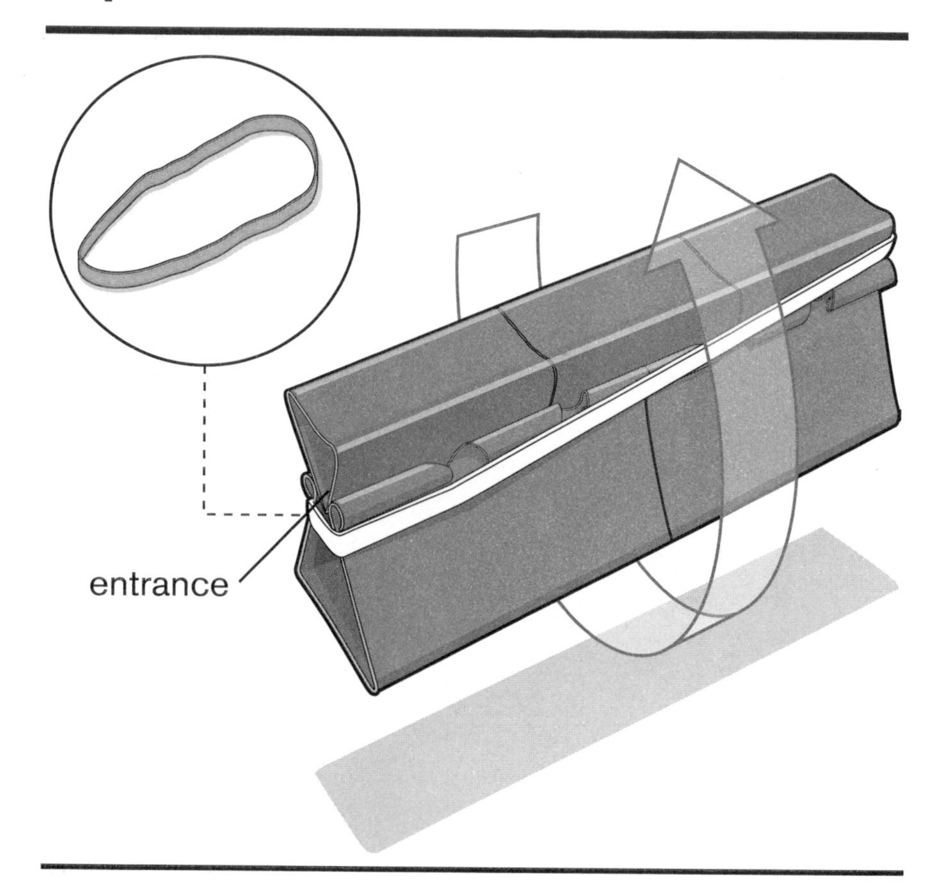


Rotate the assembly so the medium binder clips are now on top.

Loop one end of a wide and durable rubber band onto the assembly so it does not obstruct your entrance; the other end of the rubber band should cover the medium binder hole in the rear.

## Step 4

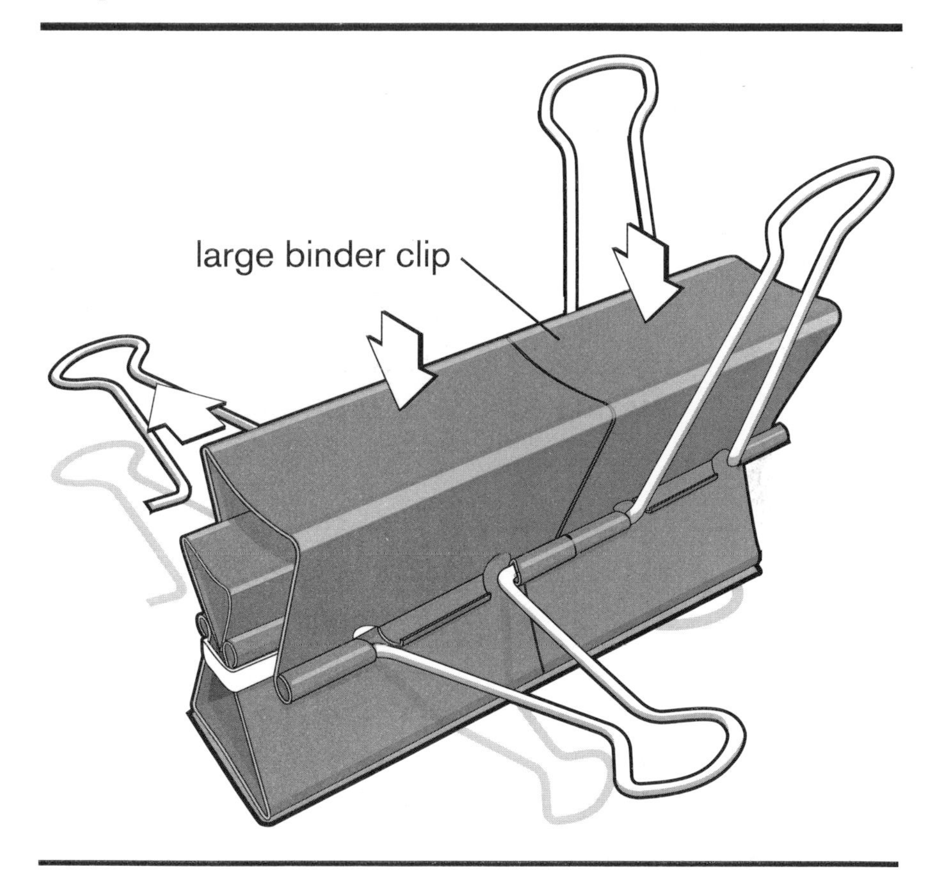


Next, clip two more large binder clips (51 mm) onto the row of medium binder clips, as shown. Once clamped on, the large binder clips should hold the rubber band in place.

Remove the metal handles from the large binder clips after you've secured them in place.

## Step 5

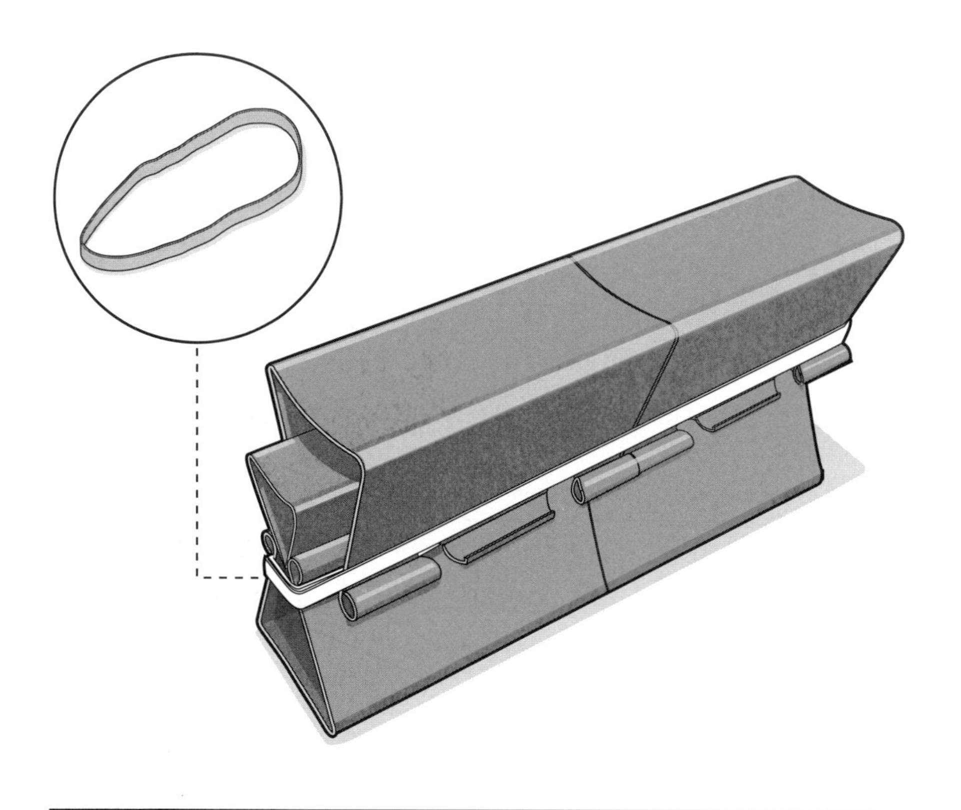

Only one more step left. Place an additional wide rubber band around the newly placed large binder clips. This will add some additional support for the Maul Gun and prevent the binder clips from sliding off.

## Step 6

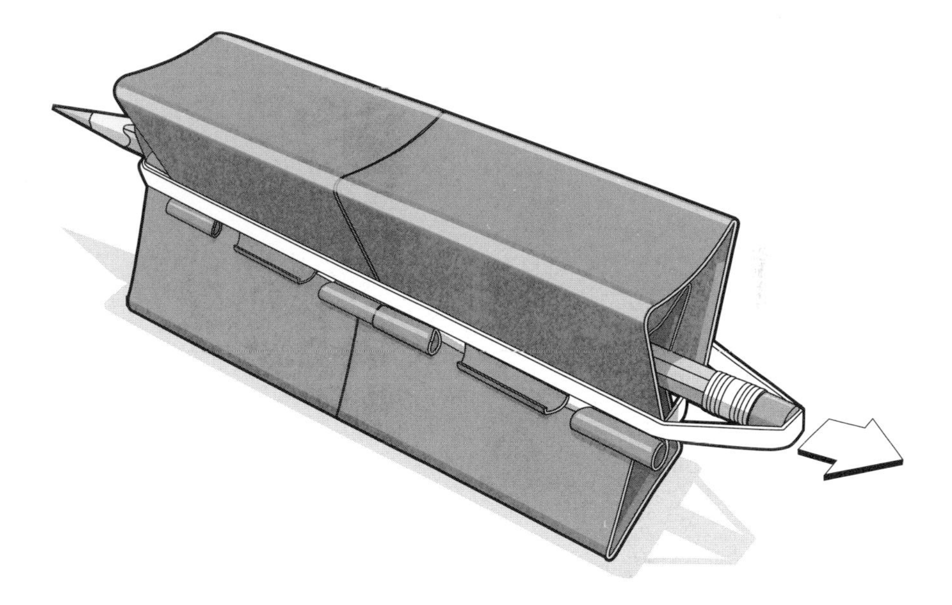

Time to fire the Maul Gun! Load a pencil or pen into the muzzle end (entrance) of the barrel. Pull the pencil and rubber band back, and once you've determined a *safe* target, release the rubber band.

A sharpened pencil is not necessary unless you are hunting aluminum cans. Unsharpened pencils, markers, highlighters, and capped pens can all be used for ammo. For safe target suggestions, visit chapter 7.

## Alternate Construction

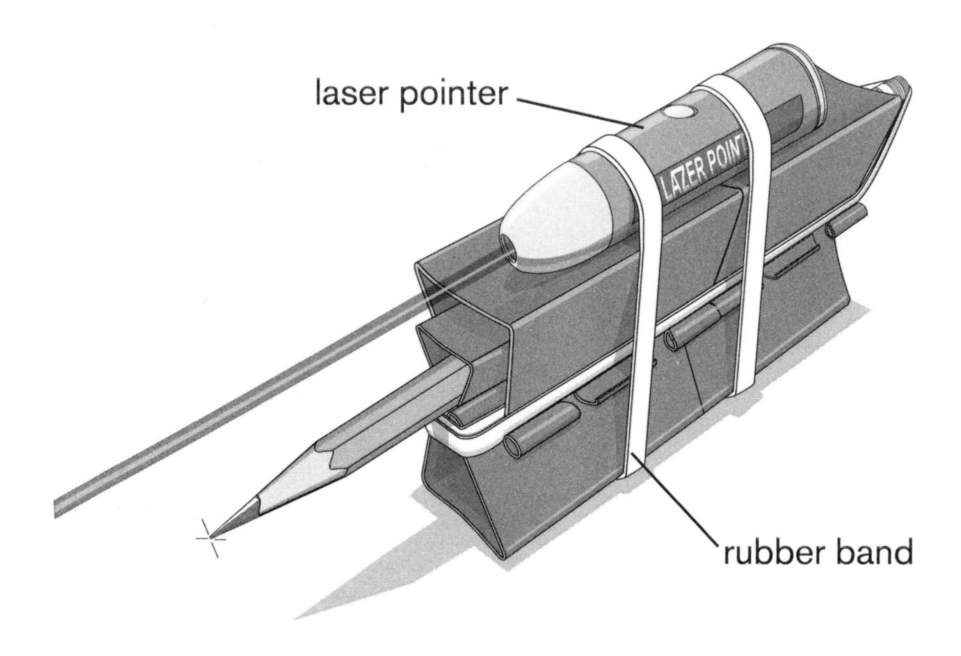


To help with accuracy and control of your Maul Gun, mount a laser pointer on top of the binder clips using rubber bands. Sight in a few times and adjust the trajectory of the red dot. This upgrade should help you narrow in on your next nonliving target.

# AIR VORTEX

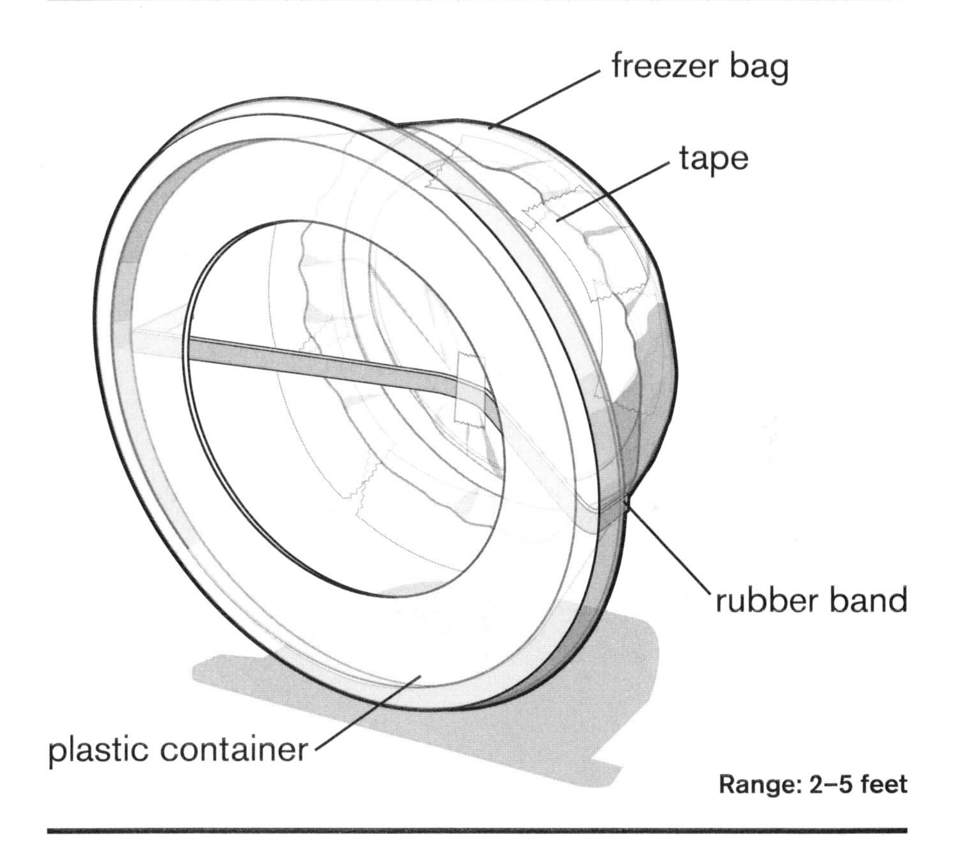


**Range: 2–5 feet**

With limited damaging powers, the Air Vortex probably can't take on much more than shattering a few helpless card houses. However, this project has huge potential for varied uses, such as extinguishing a birthday candle, blowing smoke, and even creating smoke rings. Explore these possibilities by performing controlled experiments with containers of different widths and depths.

### Supplies

1 circular plastic container with lid
1 freezer bag
1 rubber band
Transparent or packing tape

### Tools

1 drinking glass
1 marker
1 hobby knife
1 small plate

### Ammo

Air

# Step 1

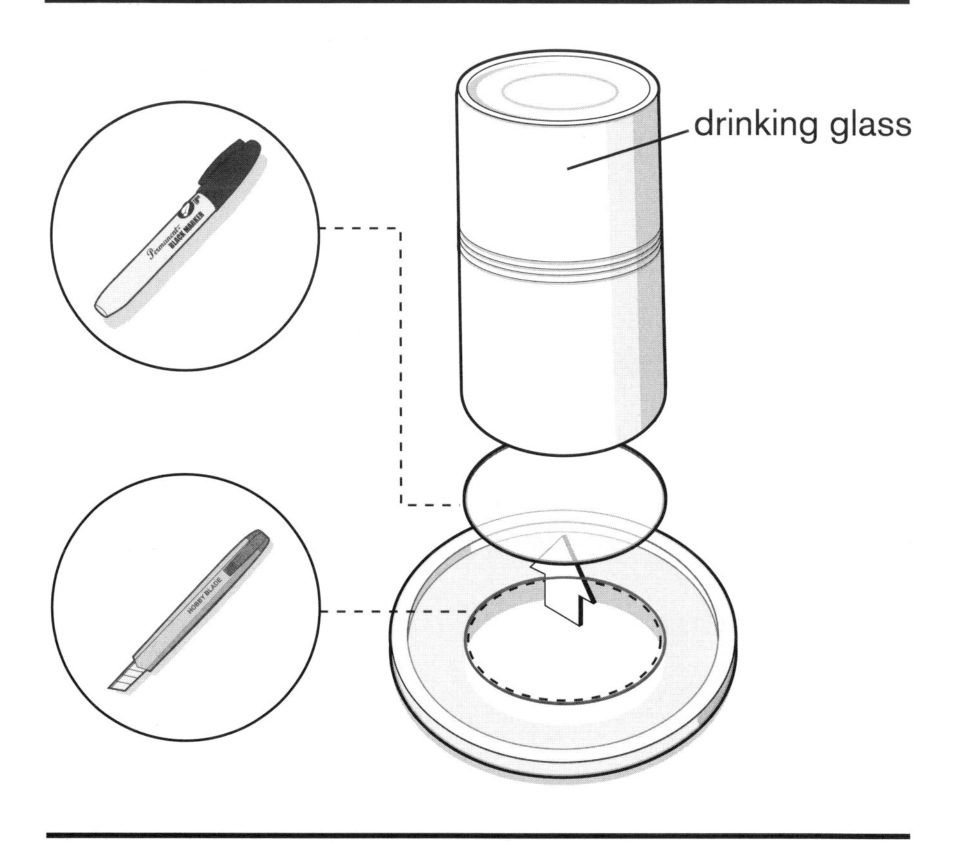


Use the drinking glass and marker to trace a circle in the center of the container lid. The area to be removed should be a few inches smaller than the container diameter. Use a hobby knife to cut the marked circle out of your container lid.

# Step 2

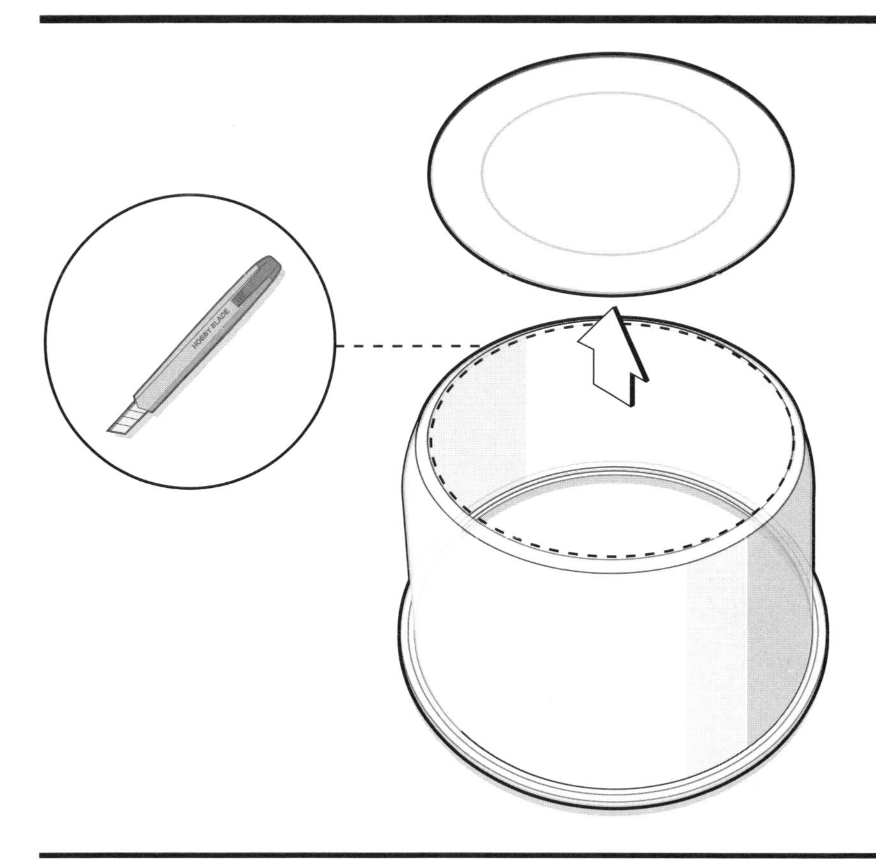


Now flip the plastic container upside down so that the bottom is facing upward. Use your hobby knife to safely cut out the bottom around the outer edge. The plastic's thickness will affect how easy it is to remove, so be careful and take your time.

You will not be using the removed container bottom, so you may recycle it.

## Step 3

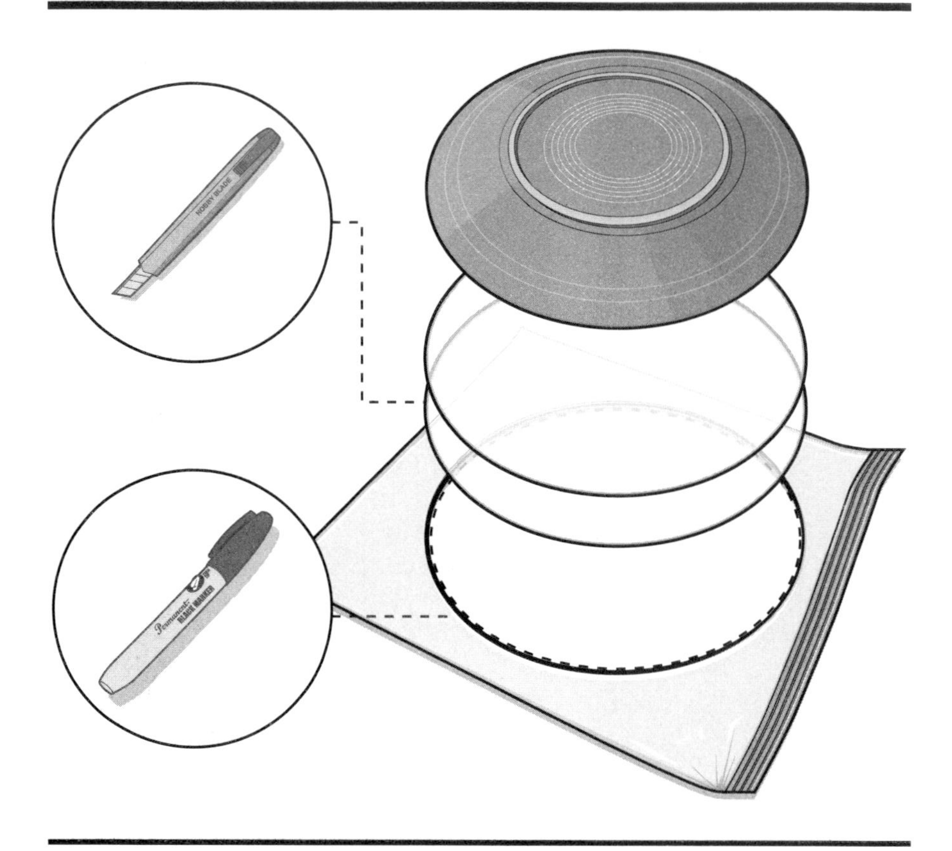

Now prepare the plastic freezer bag. Find a large freezer bag and a dinner plate for tracing. The plate's diameter should be a few inches larger than the diameter of the container you have chosen. If you have decided to build this vortex out of a larger container or cardboard box, a plastic trash bag can also be used.

Place the plate upside down on the freezer bag and trace around it with a marker. Then use scissors or a hobby knife to cut out the circle. You will need only one of the plastic circles for this project. Discard the remaining scrap.

## Step 4

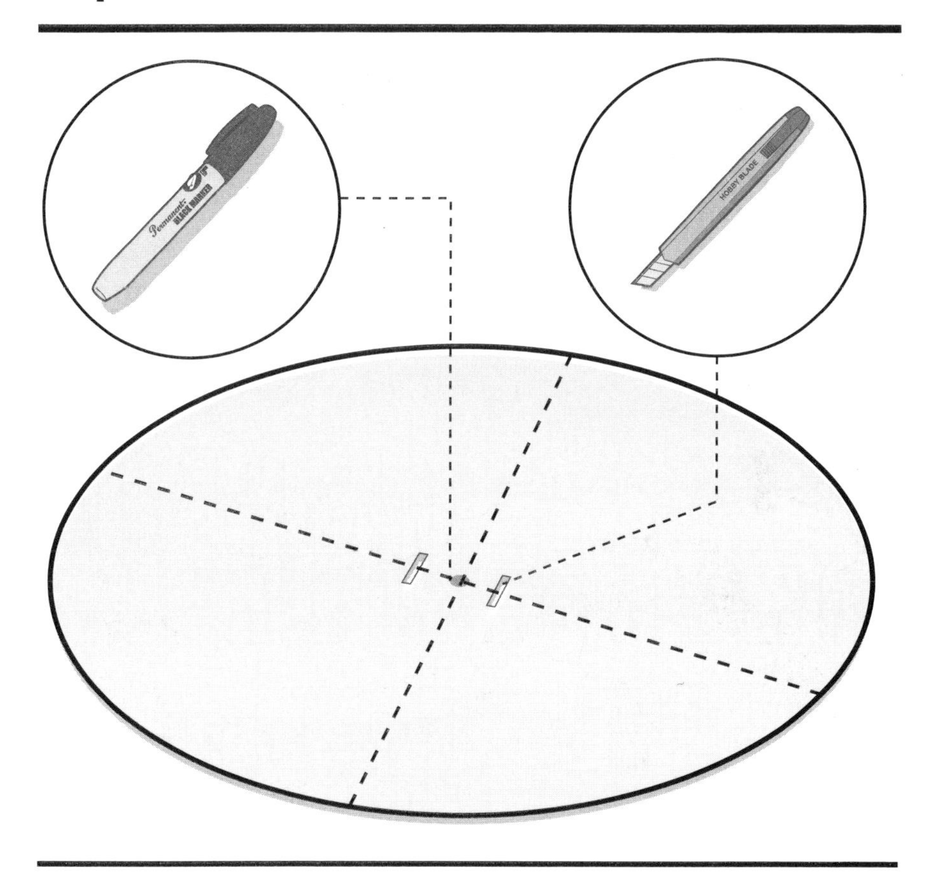

To determine the center point of the plastic circle, fold it in half, and then fold it again. Unfold the circle and use your marker to place a small dot on the point where the two folds intersect—this is the center point.

Using a hobby knife, cut two small slits into the plastic, as shown. These slits will ultimately hold a wide rubber band, so don't make the slits too large. They should be roughly the same size as the width of a wide rubber band.

# Step 5

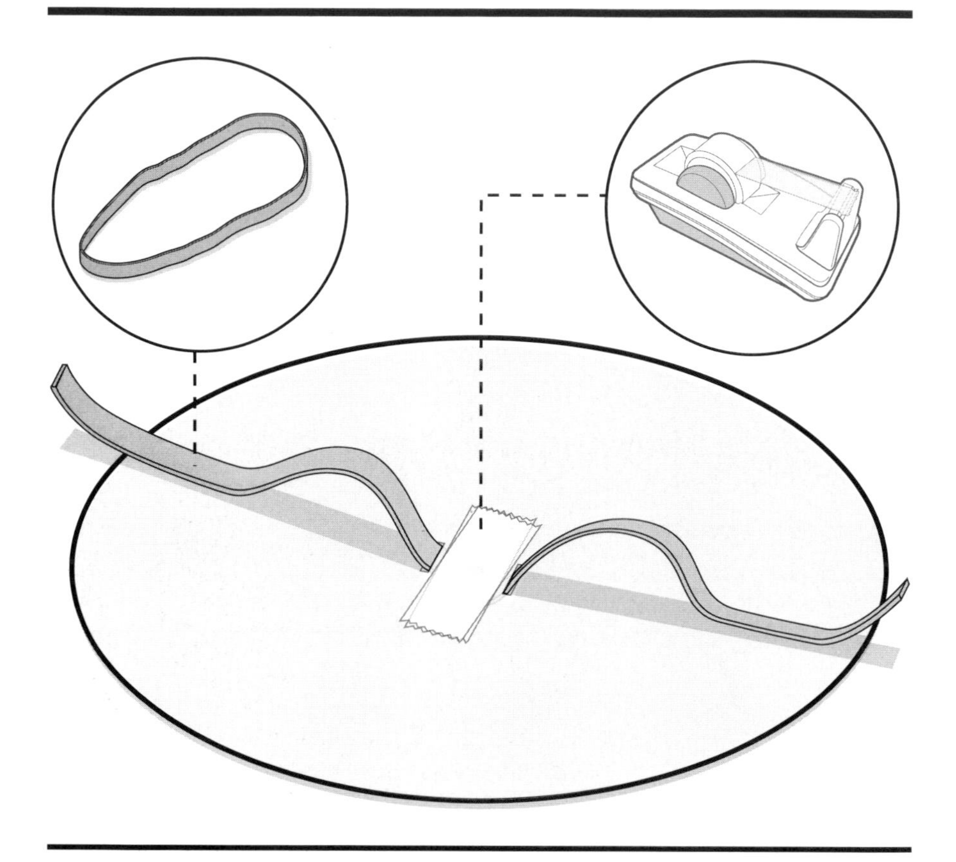

Use scissors to cut a wide rubber band. Slide the band through the slits in the plastic circle. Once you've centered the rubber band, add tape to the area between the slits to increase its strength. This added support should help prevent tearing during use.

## Step 6

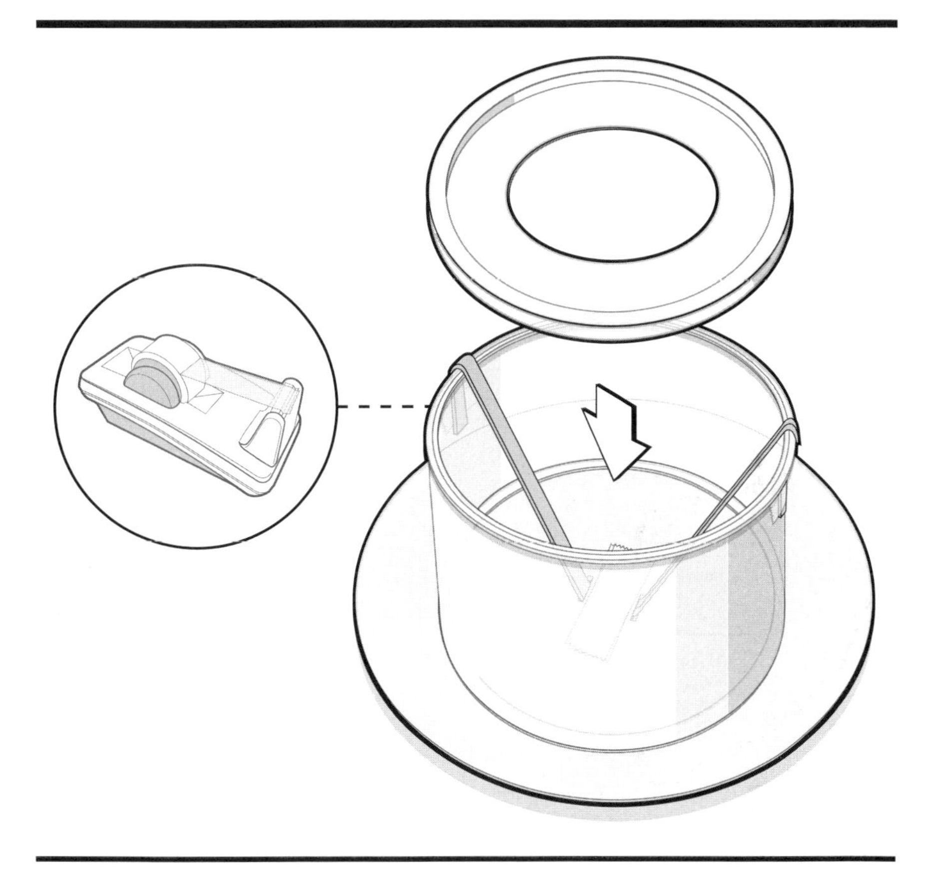

Now place the container, right side up, on top of the plastic circle, as shown in the illustration above. Pull the two ends of the rubber band over the rim of the container, parallel to each another. Tape the ends securely to the outside of the container.

Next, snap on the modified container cover. Once on, the cover should help hold the rubber band in place. But if the cover isn't holding, add tape.

# Step 7

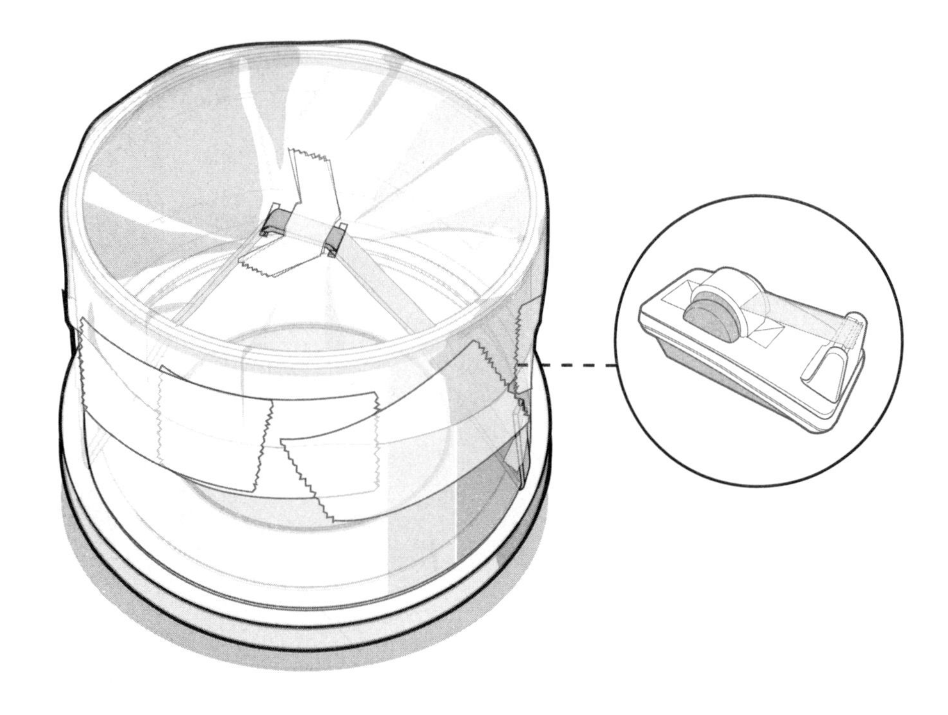

Turn the plastic container over and bend the freezer bag over the side-walls. Once the plastic is in place, go tape-crazy. When you are finished, the bond should be airtight.

## Step 8

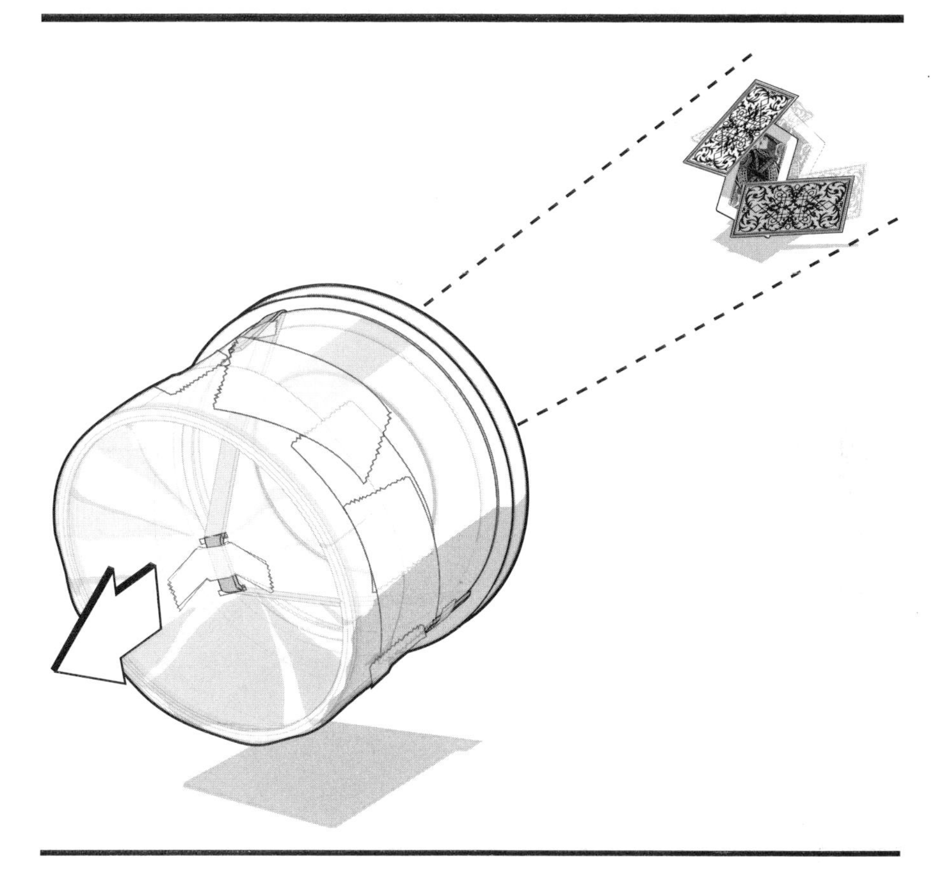

Time to destroy something! Pull back the area where the rubber band and plastic meet in the back, then release. The Air Vortex will release a gust of air directed at your target.

This is a great project that encourages experimentation and substitution of materials. Containers of different sizes, including boxes and cardboard oatmeal tubes, will produce varying results. This makes for the perfect inexpensive assignment for a group of individuals battling it out with science and physics.

The Air Vortex is completely harmless, but use the hobby knife with care.

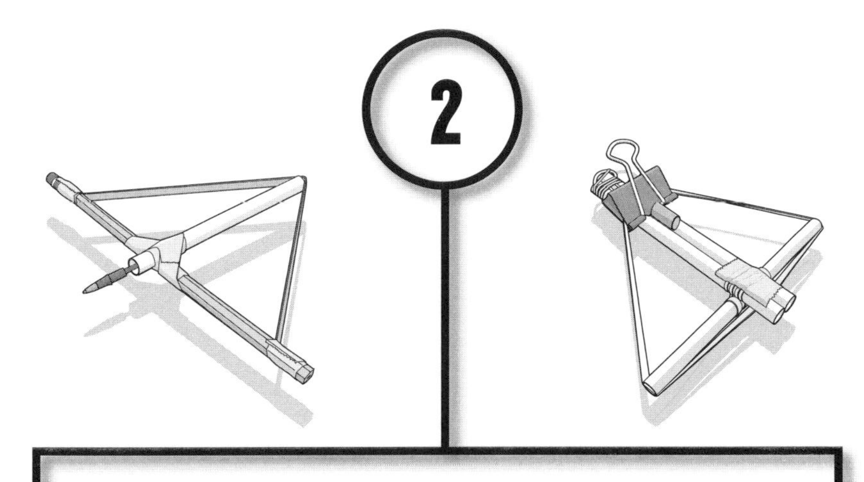

# 2

# BOWS AND SLINGSHOTS

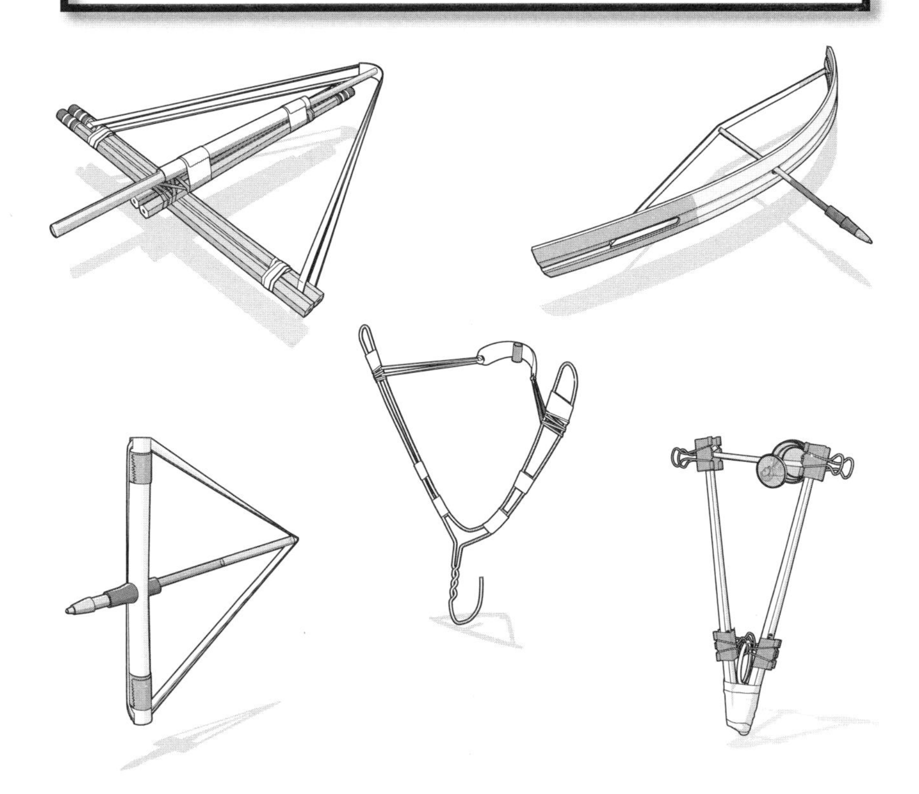

# SIMPLE CROSSBOW

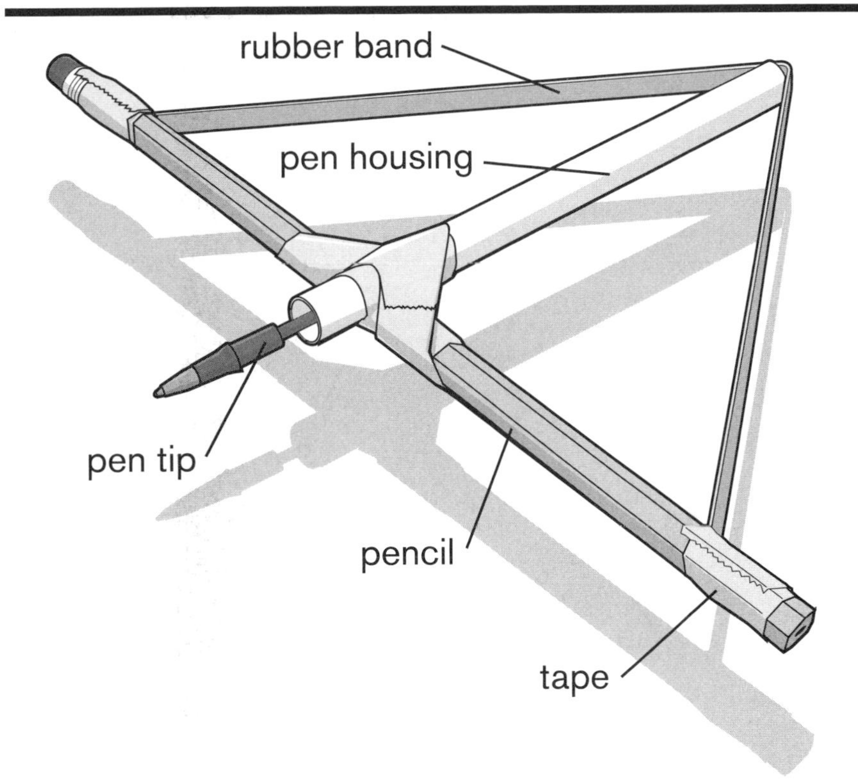


**Range: 10–30 feet**

Simple is right! This basic pistol crossbow can be assembled in seconds. Its small size and minimal parts list make it a great introductory bow. The rubber band, mounted on top of the pen housing stock, makes firing ink bolts easy and quick.

## Supplies

1 unsharpened pencil
Masking or duct tape (duct tape works best)
1 wide rubber band

## Ammo

1 pen

## Tools

Safety glasses
Pocketknife or pliers (optional)
Scissors

# Step 1

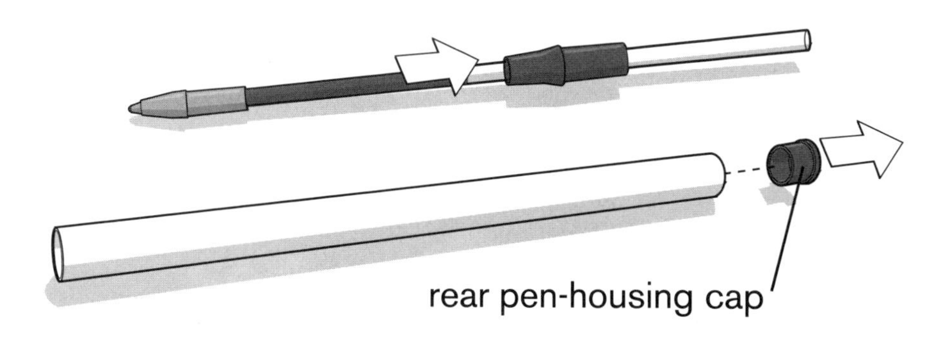


Disassemble a plastic ballpoint pen into its various parts. Depending on how the pen has been manufactured, you may need a tool to assist you with dislodging the rear pen-housing cap. A pocketknife (for cutting it off) or small pliers (for pulling it out) can both work just fine. Once the pen is disassembled, make sure the ink cartridge can pass through the pen housing. If it can't, cut an inch off your pen housing. This will allow you to pull back your ink bolt before launch.

## Step 2

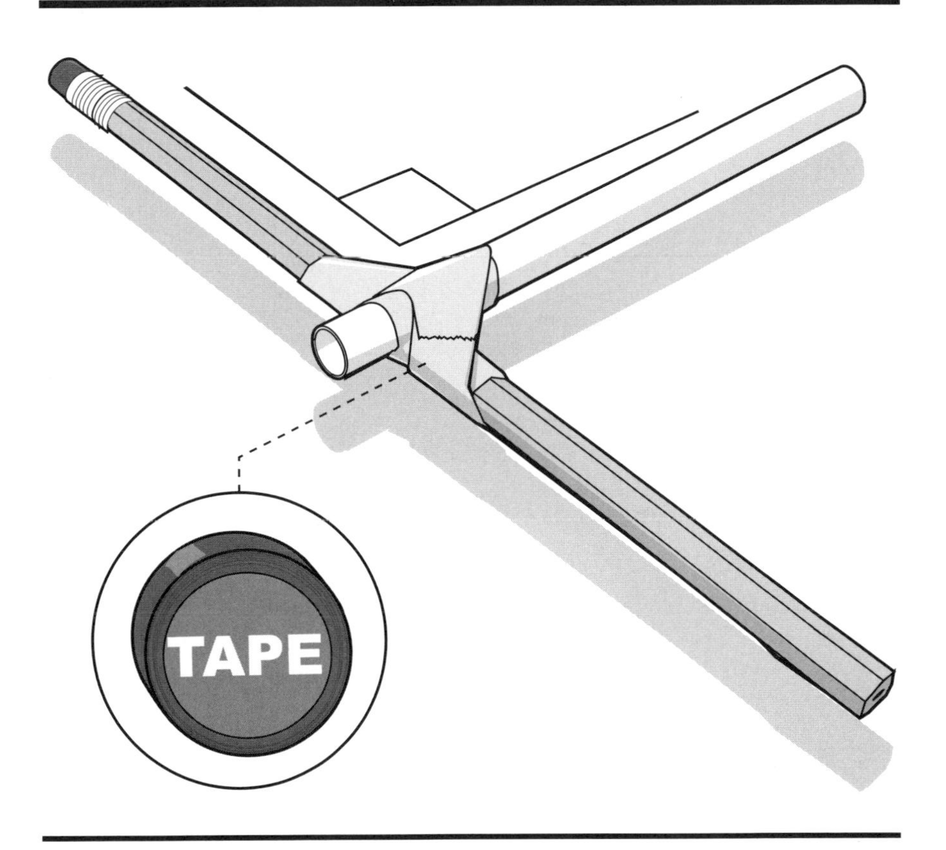


Place the hollowed-out pen housing on top of your unsharpened pencil as illustrated above. Notice that it is centered on the pencil, but only a short part of the pen housing sticks out in front. Once the pencil and pen are correctly positioned, tape the pen in place. It may take some additional taping to help center the pen housing. A 90-degree angle is ideal.

## Step 3

Cut a wide rubber band using scissors. Tape the ends of the rubber band to the ends of the pencil as shown above. You may want to tie or wrap the rubber band around the ends of the pencil before taping it in place. This will increase the bond.

Now load your pen-ink bolt into the muzzle end of the housing. Pull both the rubber band and the pen tip backwards, then release. Because of the light weight of your arrow, accuracy will be limited. Always operate with safety glasses.

# CLIP CROSSBOW

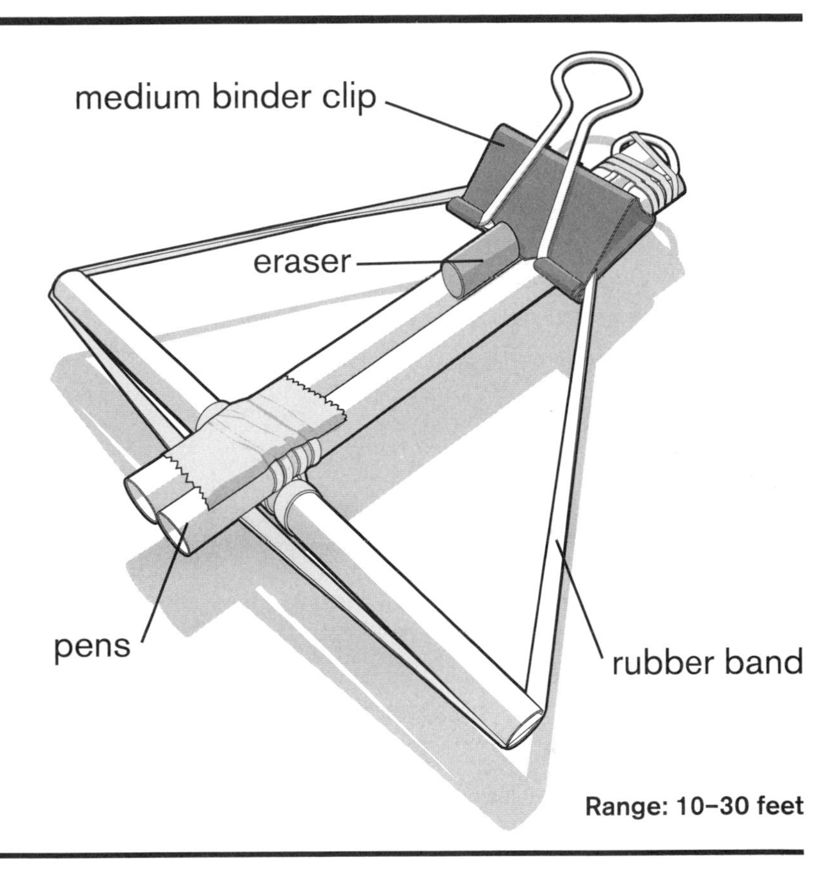


The Clip Crossbow is a great universal MiniWeapon with endless ammo possibilities. It's designed to hold the bow in a fully drawn position until you are ready to release the binder-clip trigger—just like a real crossbow! Once it's complete, you'll enjoy its single-hand operation.

## Supplies

3 pens
4 thin rubber bands
1 medium binder clip (32 mm)
1 wide rubber band
Tape (any kind)

## Tools

Safety glasses
Hobby knife
Pliers (optional)

## Ammo

1 eraser

## Step 1

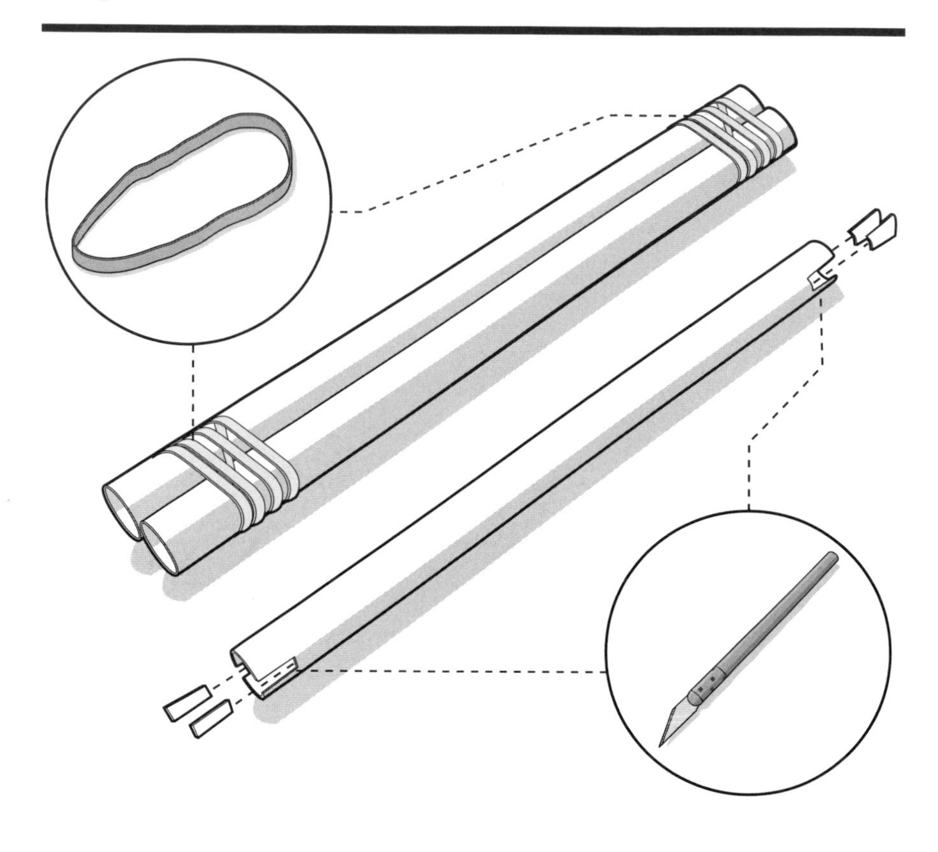

First, round up three plastic ballpoint pens. Using some elbow grease and possibly a tool or two, remove all the contents from the pen housings.

Use two of your thin rubber bands to bind two of the pen housings together. On the third pen housing, cut two small notches opposite one another on both ends, as shown. These notches will hold one wide rubber band.

# Step 2

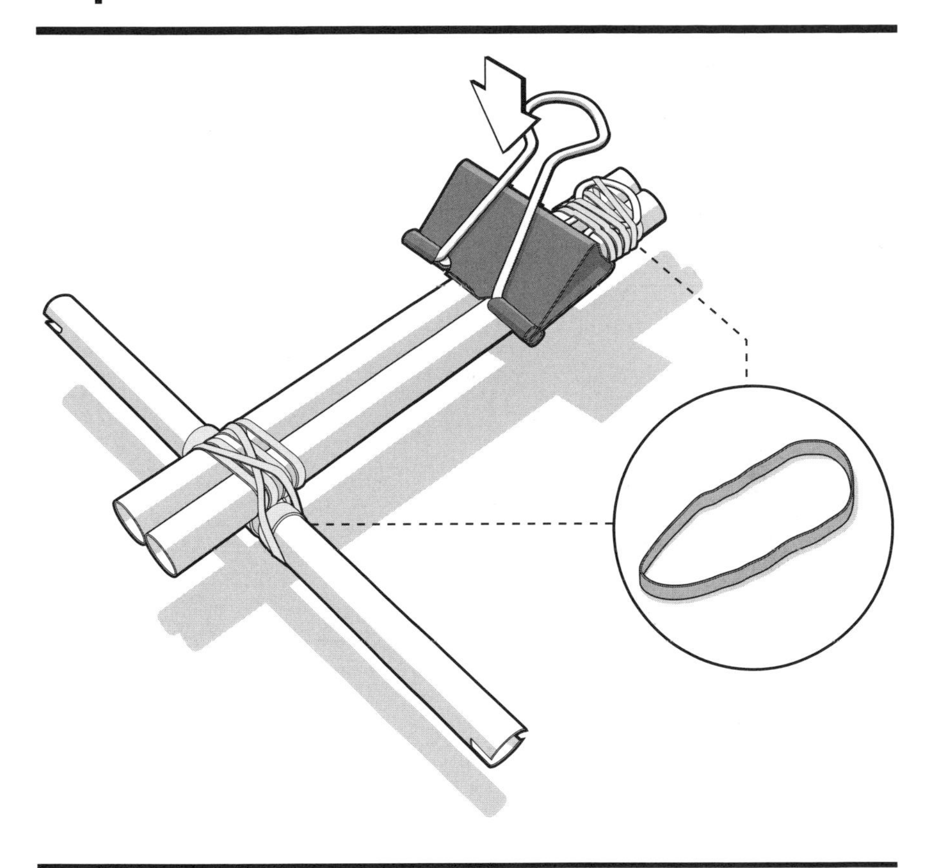

Now it's time for some light assembly. Start with the crossbow arm. Rubber band the third pen housing (with the notches) to the bottom of the double frame. You're looking for a 90-degree angle. Also, the notches on the third housing should be oriented parallel to the double frame, as shown.

Next, install a medium binder clip to the back of the crossbow: Place the clip on the top of the double-pen frame, with clip end facing forward, and use a rubber band to hold it in place. Only the bottom metal arm should be rubber banded—the top metal handle should remain free for loading and unloading the elastic band.

## Step 3

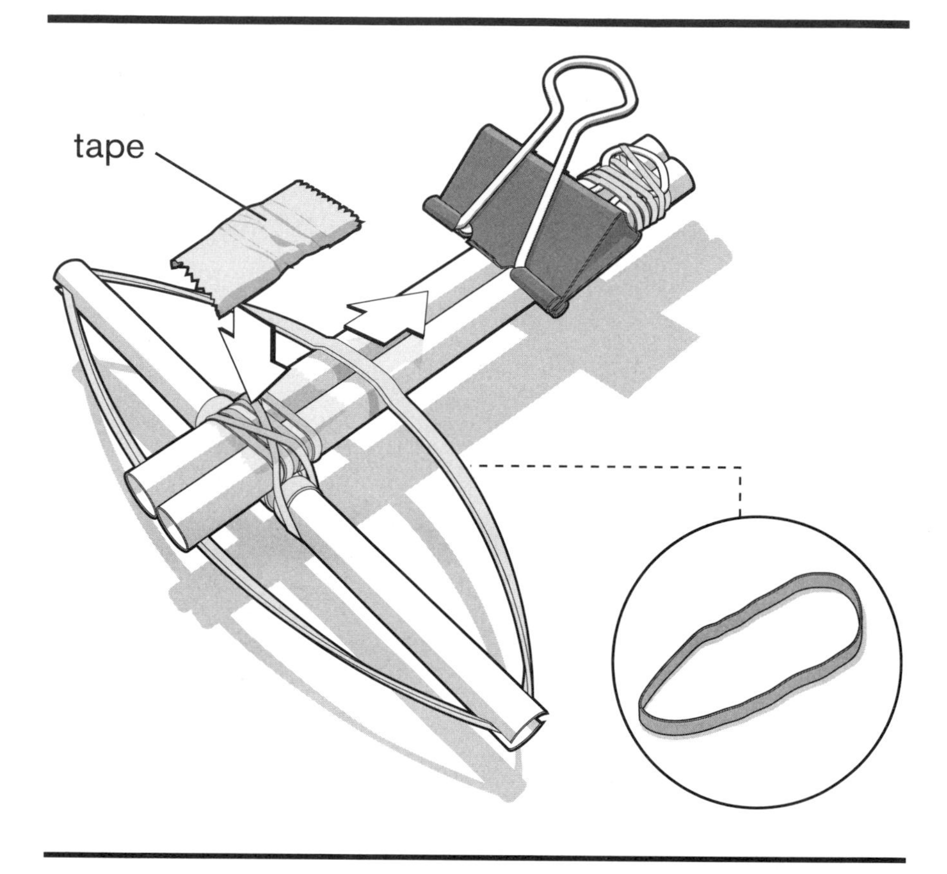


Loop one wide rubber band through the pen housing groves you cut earlier. Then pull back the band and clip it into the binder clip. Cover the wrapped rubber band on the barrel with a single piece of tape to help smooth out the trajectory for your ammo.

Your Clip Crossbow is now in the fully drawn position until you release the binder clip. Place the recommended eraser ammo in front of the clip before pressing the metal handle down. (See the final illustration on page 59 for placement.)

Always wear safety glasses when operating your Clip Crossbow.

# #2 CROSSBOW

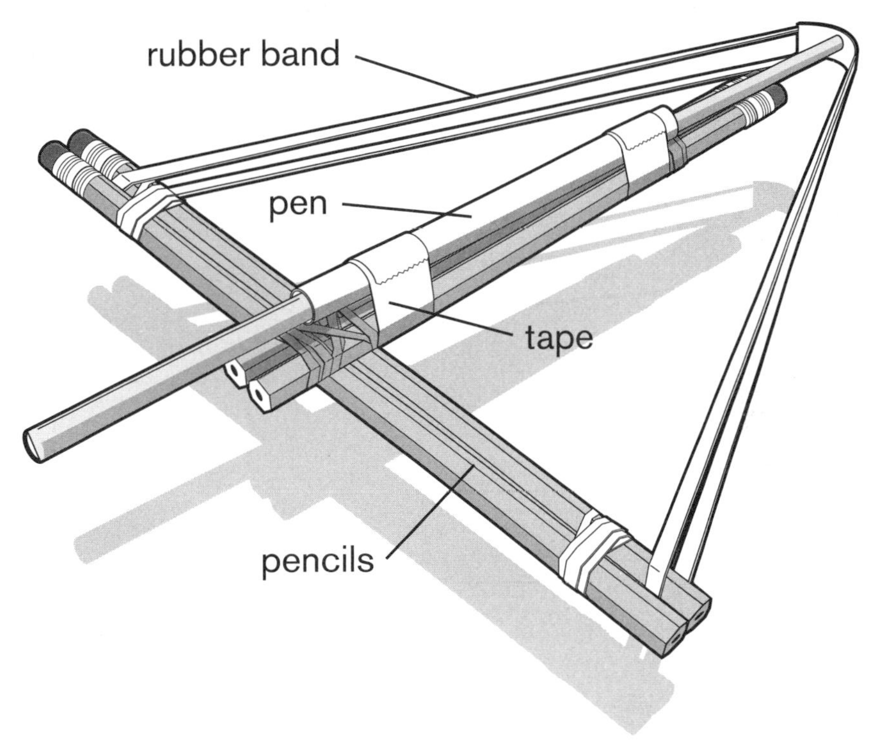


**Range: 20–40 feet**

The #2 Crossbow is a larger variant of the other bows found in this book. With a structurally solid design and double elastic power, it is equipped to fire large, realistic skewer arrows. It sports a pen-housing barrel that helps with both accuracy and control.

## Supplies

4 pencils
5 or 6 thin rubber bands
1 pen
Tape (any kind)
2 wide rubber bands

## Tools

Safety glasses
Pocketknife (optional)
Pliers (optional)

## Ammo

1+ wooden skewers (or 3/16-inch dowels)

## Step 1

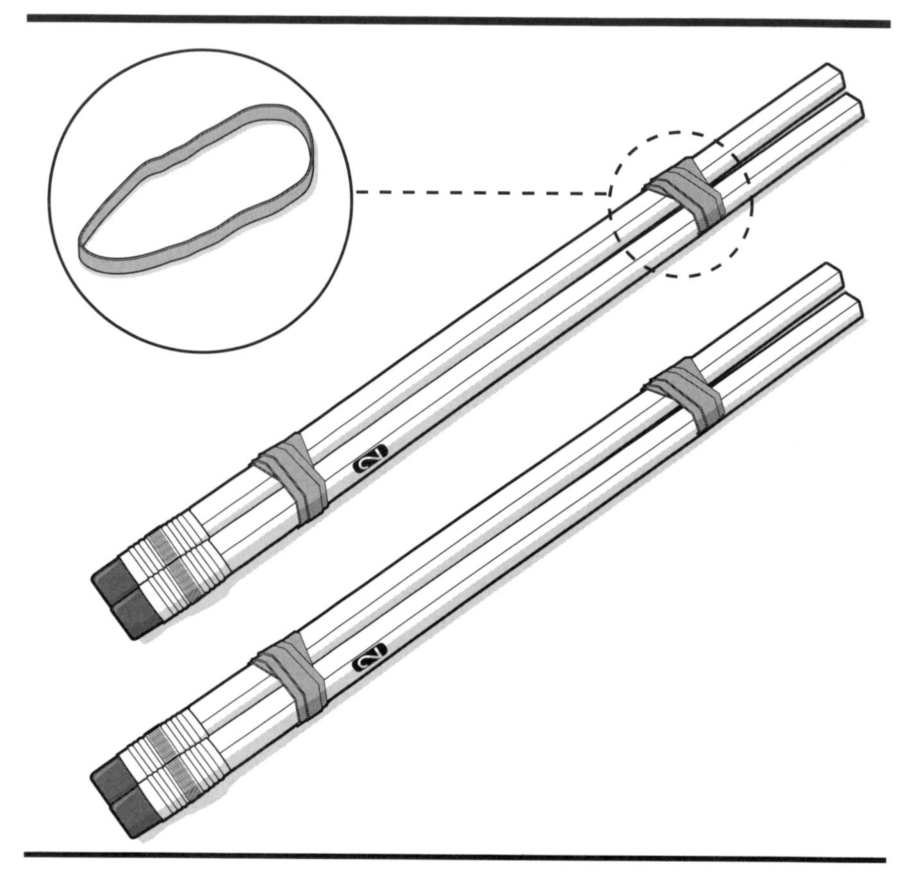

Using four thin rubber bands, assemble two pairs of unsharpened wooden pencils. Both sets should be identical and tightly secured.

## Step 2

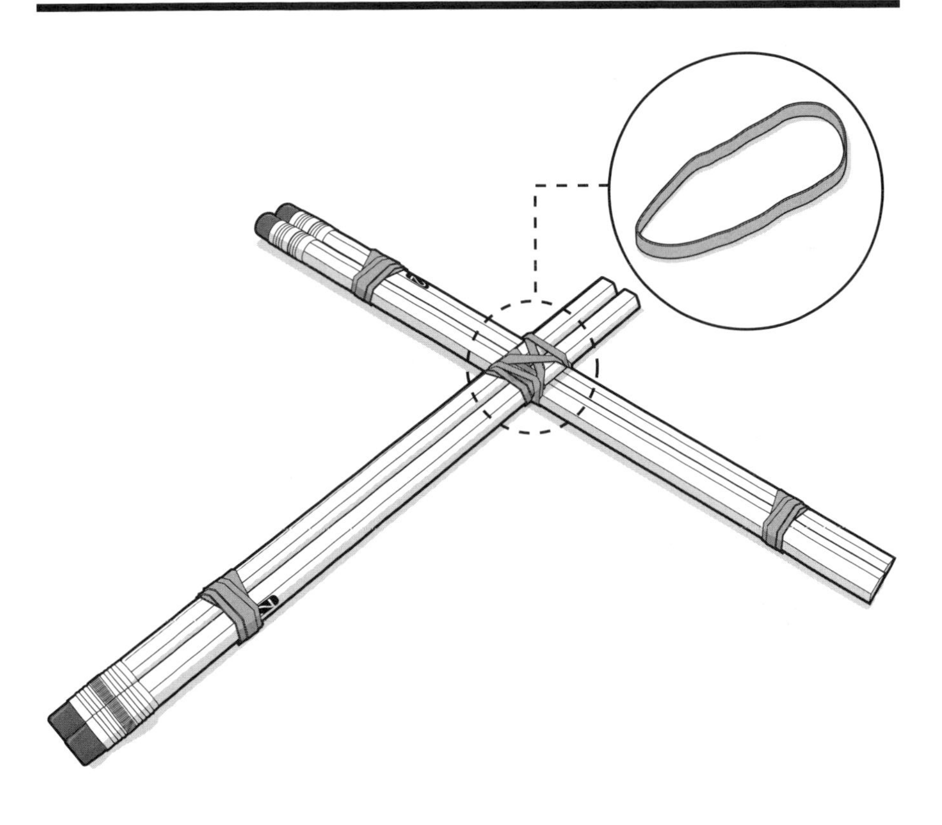

Crisscross the two sets of pencils. Center one of the pairs on top of the other towards a selected end. This end will ultimately be the front of your #2 Crossbow. While holding the pairs in place, use one or two thin rubber bands to fasten the frames together.

## Step 3

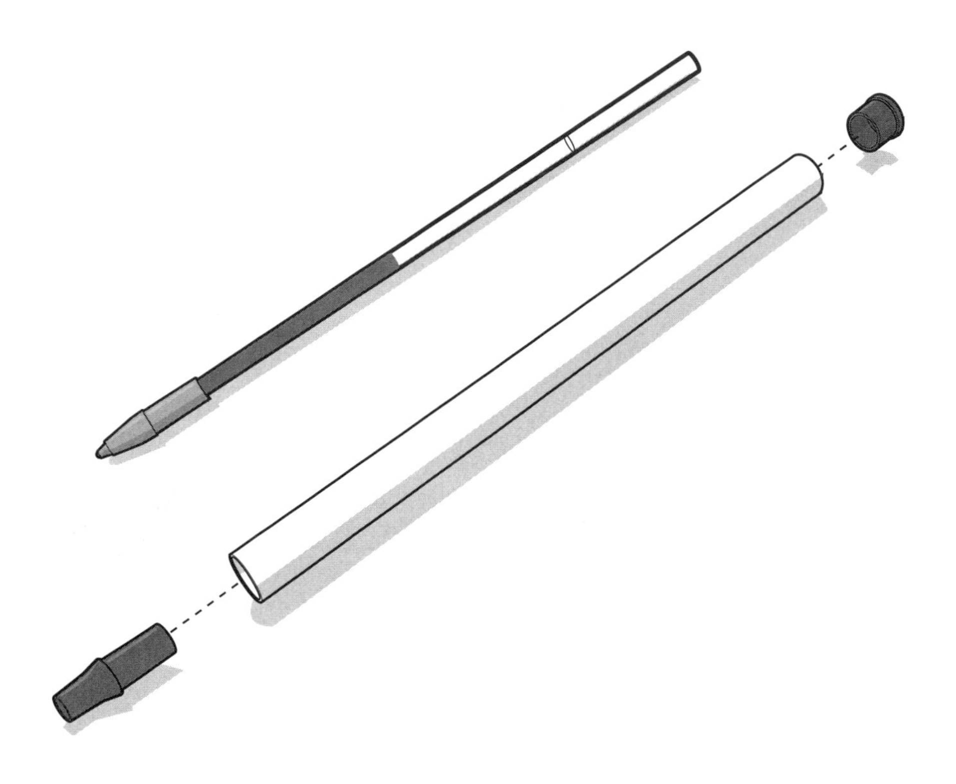

Disassemble a plastic ballpoint pen. You may need a pocketknife or pliers when removing the rear pen cap. The hollowed-out pen housing will be used for your crossbow barrel. Discard all the other pen contents (or save them for the Ruler Bow on page 71).

## Step 4

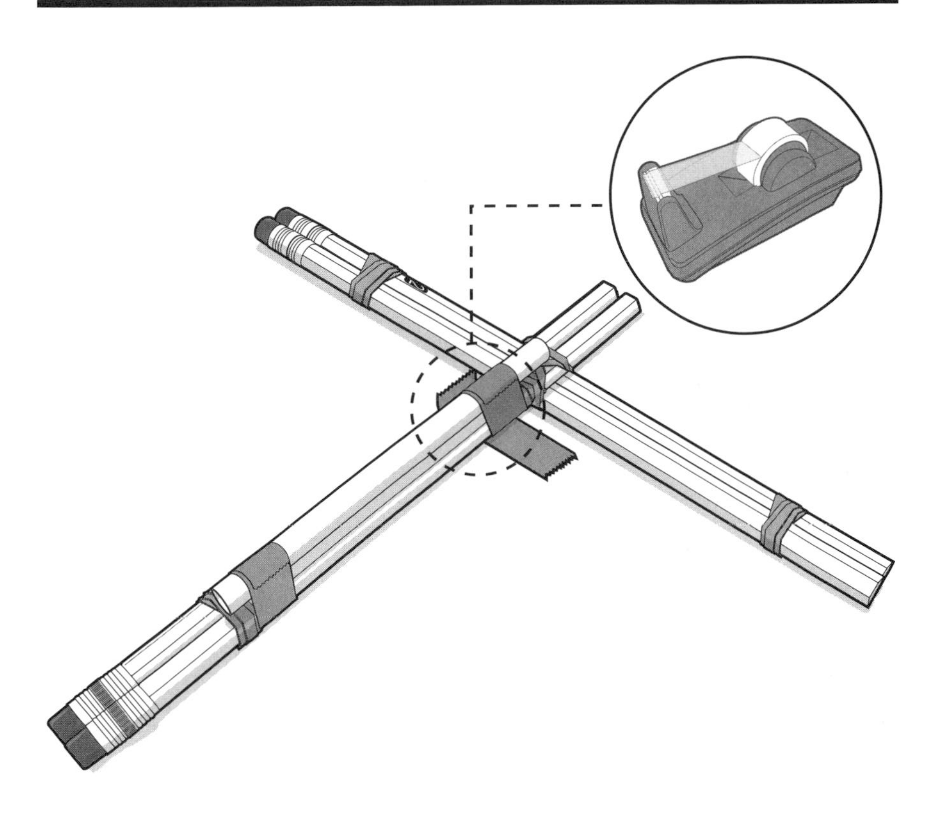

Position the pen housing on top of the pencils, as shown, then secure with tape. It is important that the pen housing sits on top of the rubber bands and that the barrel is not obstructed by them.

## Step 5

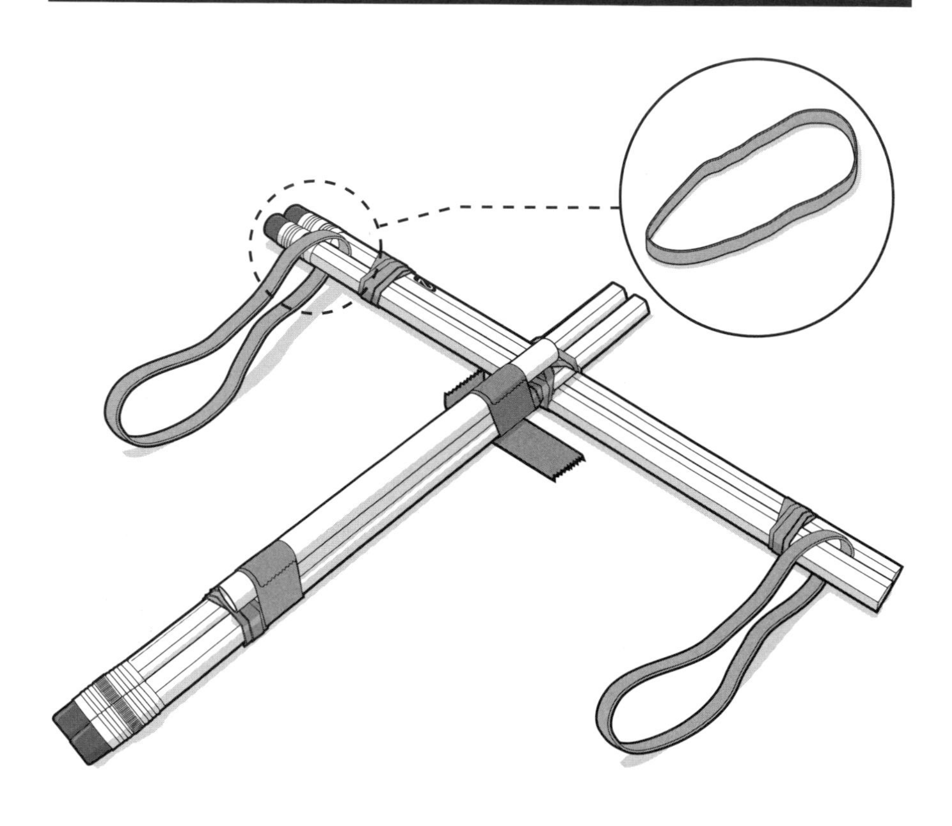

Slide two wide rubber bands between the tightly secured pencils ends. The pencils should lock the bands in place, but if they don't, add additional thin rubber bands on the ends. The wide rubber bands will ultimately provide you with your elastic firepower.

## Step 6

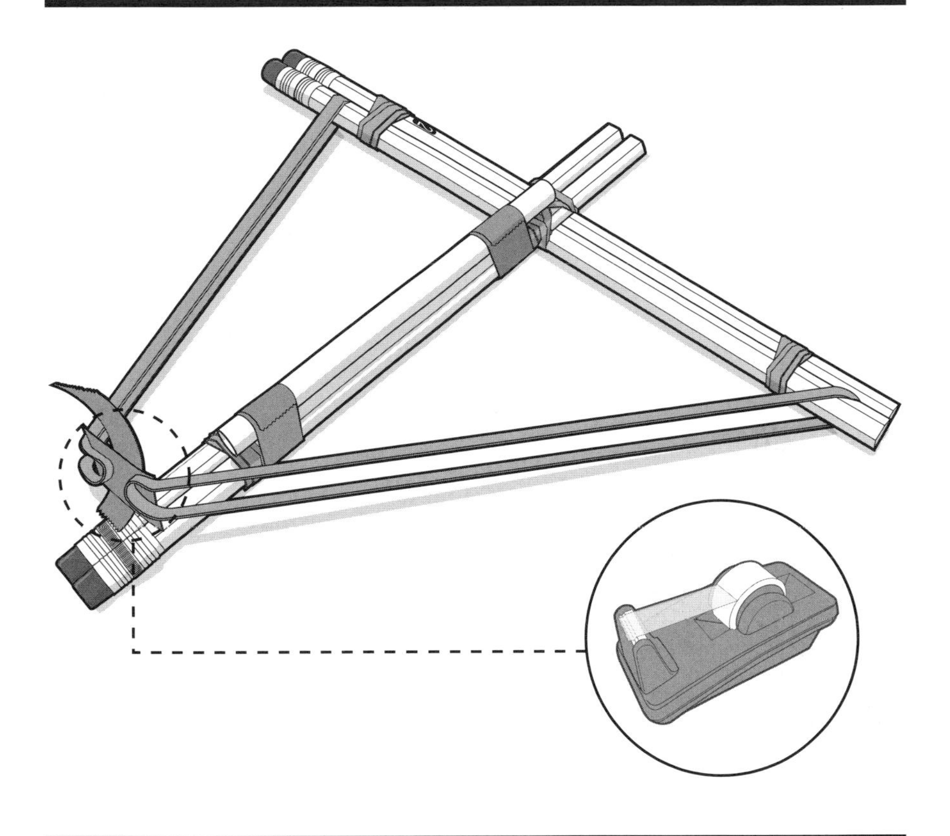

Bring both ends of the rubber bands together and attach them using strong tape. As you secure the bands, try to create a small ammunition pouch with the tape. It is possible you may need several pieces of tape to fasten the bands together securely. Pull the assembly back a few times with your finger to test.

## Step 7

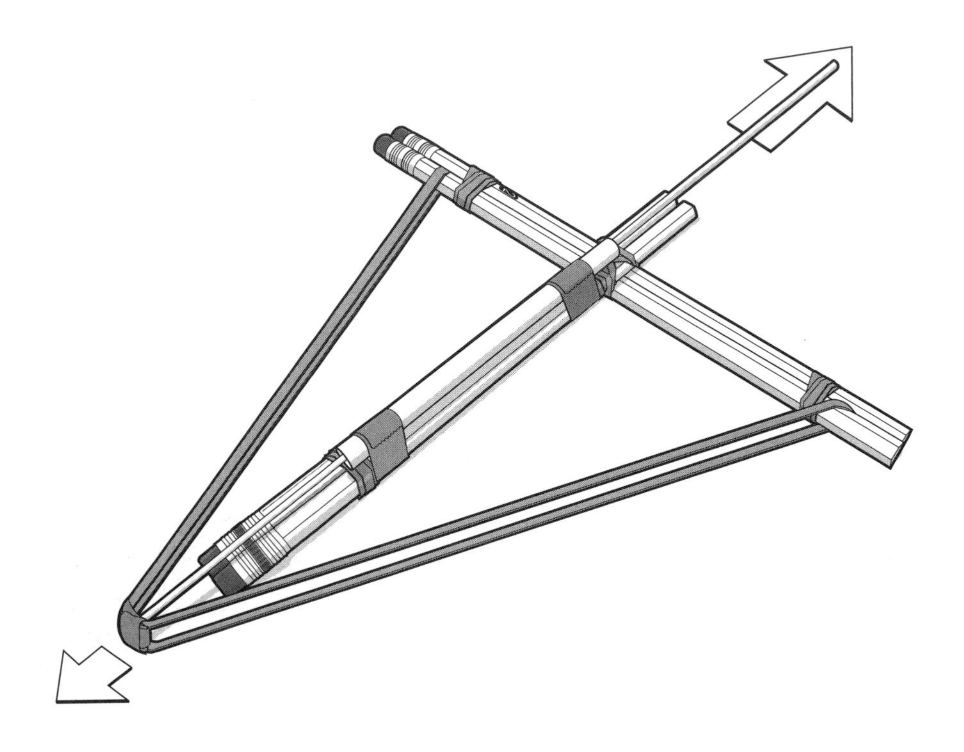

Your #2 Crossbow is now complete. Slide one wooden skewer (used for cooking) or a 3/16-inch dowel into the pen housing. Gripping the wooden arrow and the rubber bands, pull back and aim your crossbow launcher. Release and watch it fly!

Always operate your crossbow safely. Watch out for spectators and never aim the shooter at anyone. Wooden skewers usually have pointed tips, which can make them very dangerous. Styrofoam targets are ideal, but you should never place them in front of a breakable backdrop just in case you miss your target. Do not use the #2 Crossbow if any of its rubber bands show signs of wear.

# RULER BOW

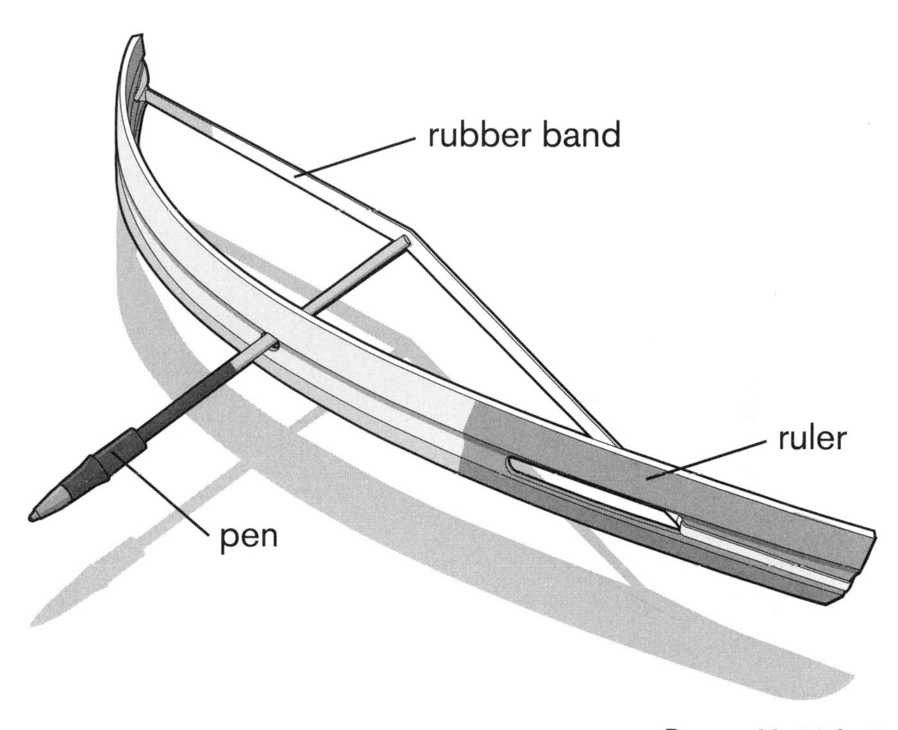


**Range: 20–40 feet**

Every office has a modern-day Robin Hood, someone who steals office supplies from the rich and gives them to the poor. So if you want to play the part, you must arm yourself with the necessary archery hardware. This Ruler Bow is one of the most basic launchers known to humankind. In fact, it's not far off from the shooting stapler or spitballs. Despite its simplicity, it fires a great distance and is quite accurate.

## Supplies

1 rubber band
1 plastic ruler

## Ammo

1 pen

## Tools

Safety glasses
Scissors

## Step 1

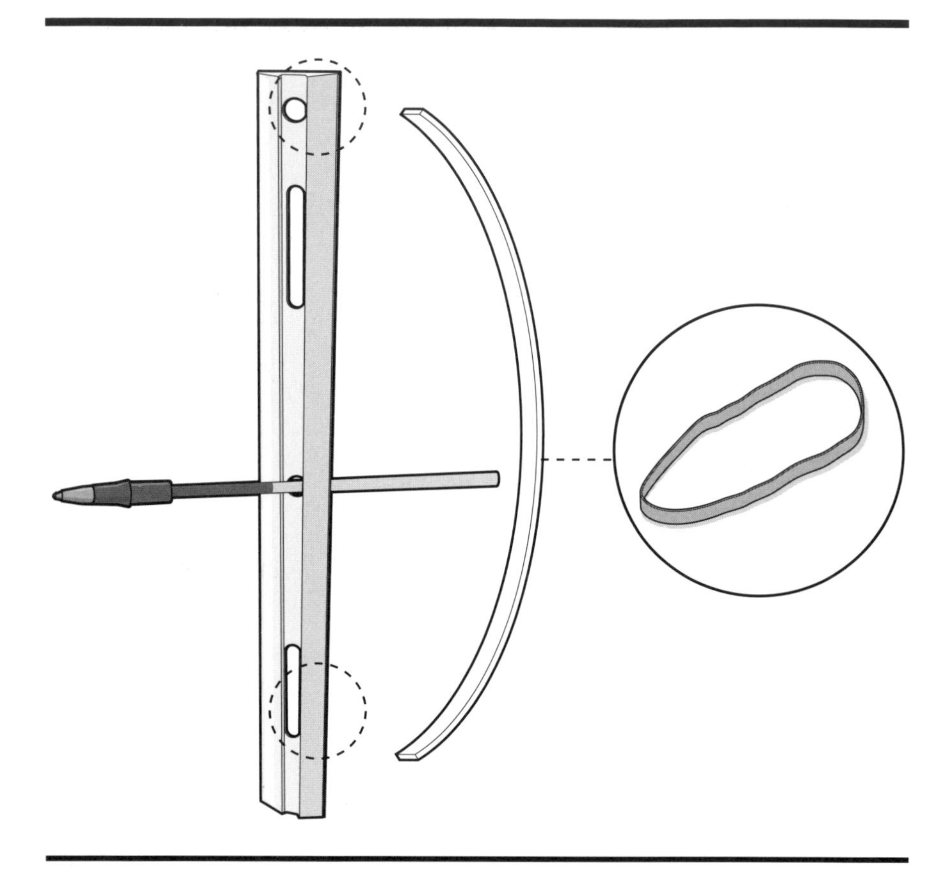

First, use scissors to sever a thick rubber band with one cut. Tie the rubber band to both ends of a plastic ruler through the factory holes already molded into it. A double or single knot will work. (Metal or wooden rulers will also work for this project, but you may have to drill the holes that are standard on most plastic rulers.

Disassemble a plastic ballpoint pen. The interior pen tip will be your ammo. (Save the housing for other MiniWeapons projects.)

Now you're ready for the hunt. Load the pen-tip arrow into the center hole of the ruler, pull back the cartridge with the rubber band, and then release your ink missile by letting go of the rubber band.

# BOW-AND-ARROW PEN

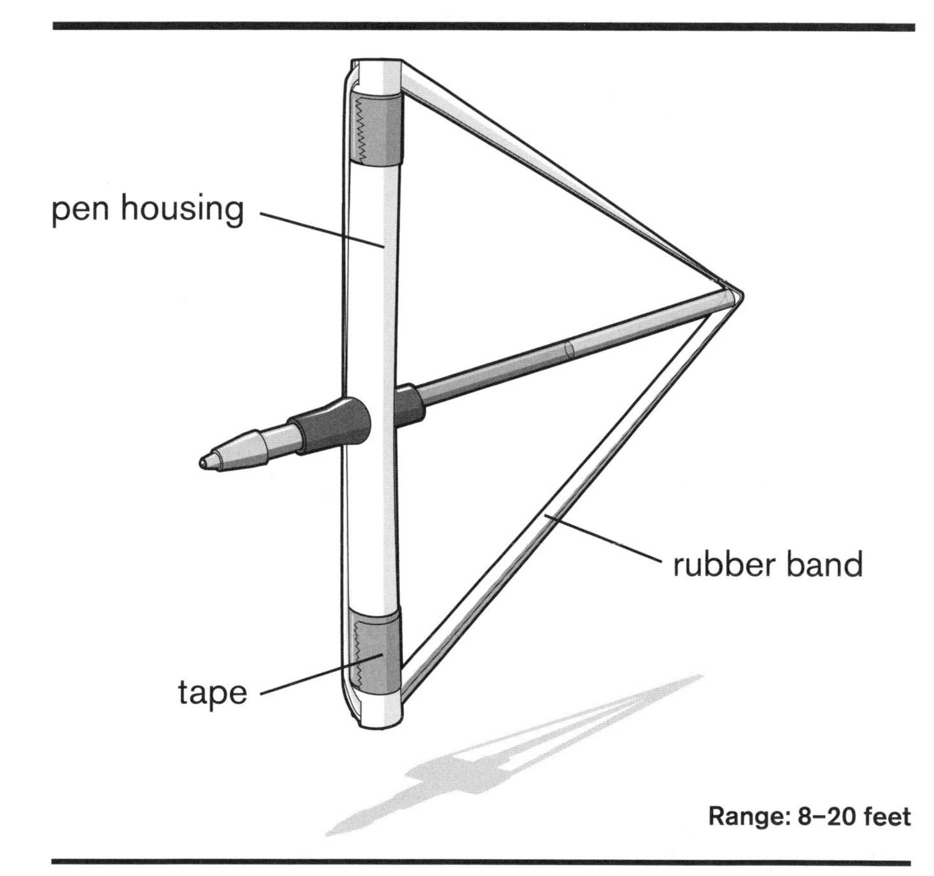


**Range: 8–20 feet**

Even Rambo needed a hobby. He mastered archery, and this sport can quickly occupy your time as well, while you perfect your mechanical principles. A customized plastic bow is perfect for long-distance competitions and target practice. Use the print-out targets in the back of the book to determine which of your fellow archers is the true Robin Hood.

## Supplies

1 wide rubber band
Duct tape

## Tools

Safety glasses
Hobby knife
Pliers (optional)

## Ammo

1 pen

## Step 1

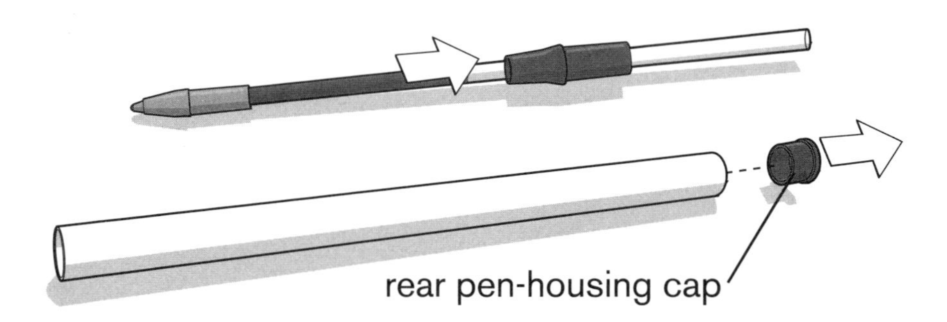


Disassemble a plastic ballpoint pen into its various parts. Unless you chew on your pens, you may need a tool to dislodge the rear pen cap. A hobby knife or small pliers should be sufficient. Lay out all the pen components and, unlike in previous builds, do not discard anything.

## Step 2

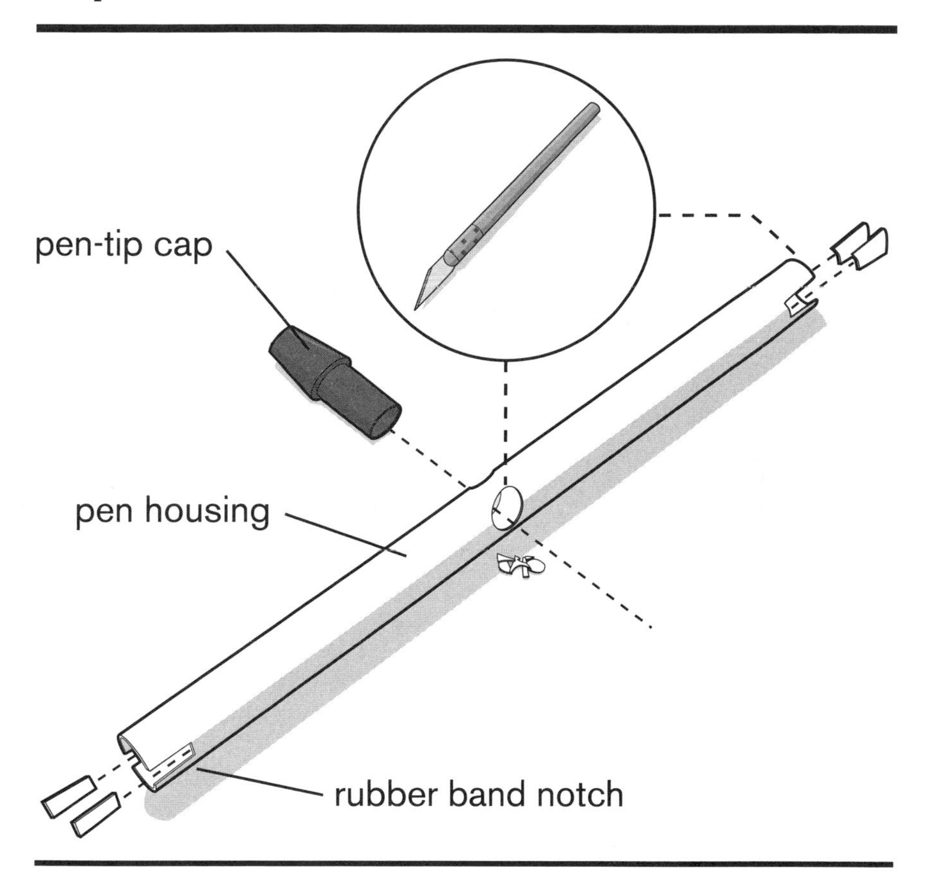


Using a hobby knife, cut two circular openings in the center of your pen housing directly across from each other. Your pen-tip cap should be able to fit snugly into the holes. Do not cut the holes too big—this will weaken the pen housing and could cause frame buckling later.

Next, cut two notches from each end of the housing. Notches should be aligned with the circular holes that you cut out for the pen tip, as shown.

## Step 3

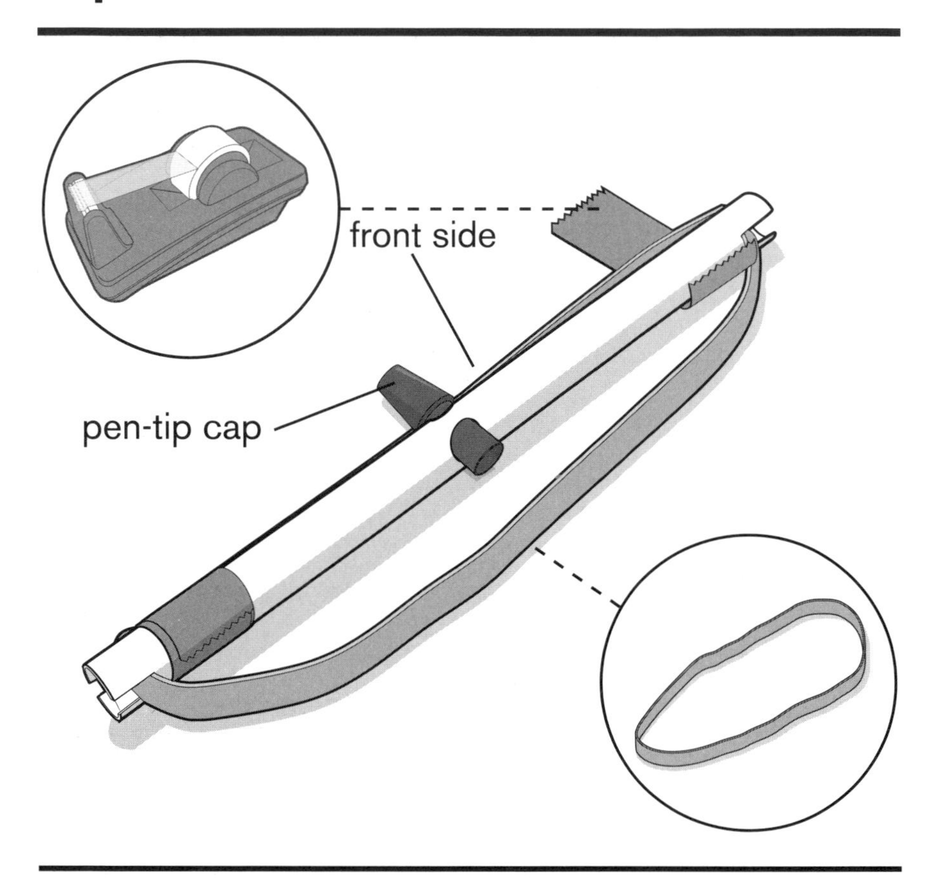


Insert a wide rubber band into the end notches you created in the pen housing. Secure the rubber band with tape just next to each notch, but only tape the rubber band to the front side of the launcher. The back half of the band should not be taped in order to allow the elastic to bow.

Now insert the pen-tip cap into the holes in the pen housing. This cap will support your ink arrows during firing. If the cap does not fit snugly, add tape for additional support. Move the rubber band up or down so it does not obstruct the hole in the pen-tip cap. You can tape it in place if you'd like.

## Step 4

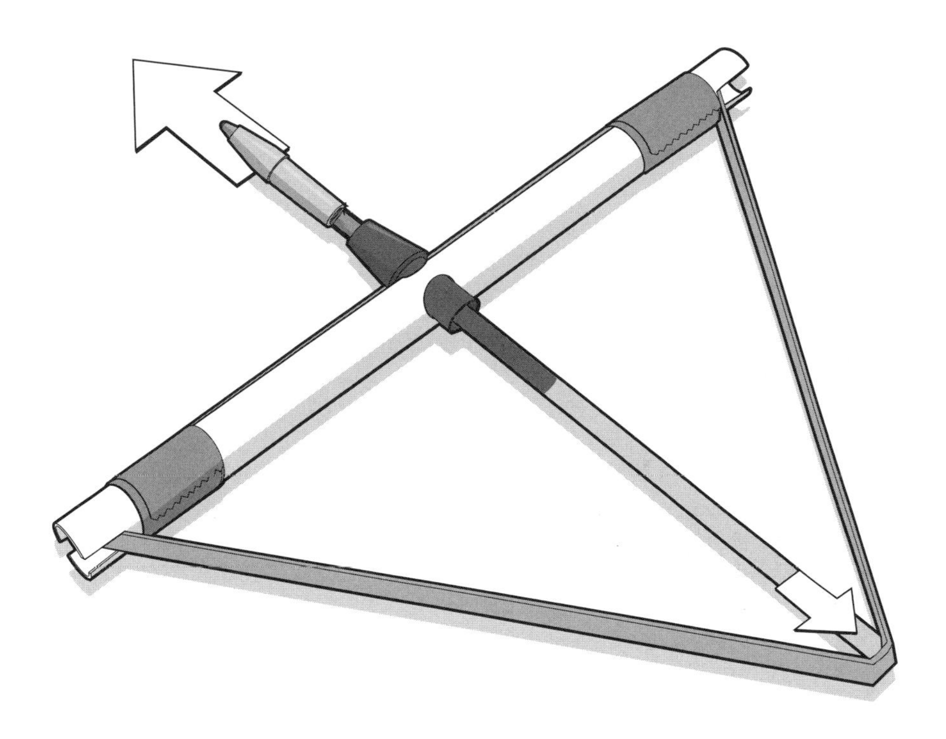

The Bow-and-Arrow Pen is now complete—time to fire it! Slide the ink cartridge through the pen-tip cap and grasp both the rubber band and the end of the ink cartridge to fire. Once the bow is drawn, release the rubber band and watch your ink arrow fly.

Remember that this homemade projectile launcher is capable of unfortunate malfunctions and misfires. Before firing, create a controlled shooting range to safely operate your weapon.

## Alternate Construction

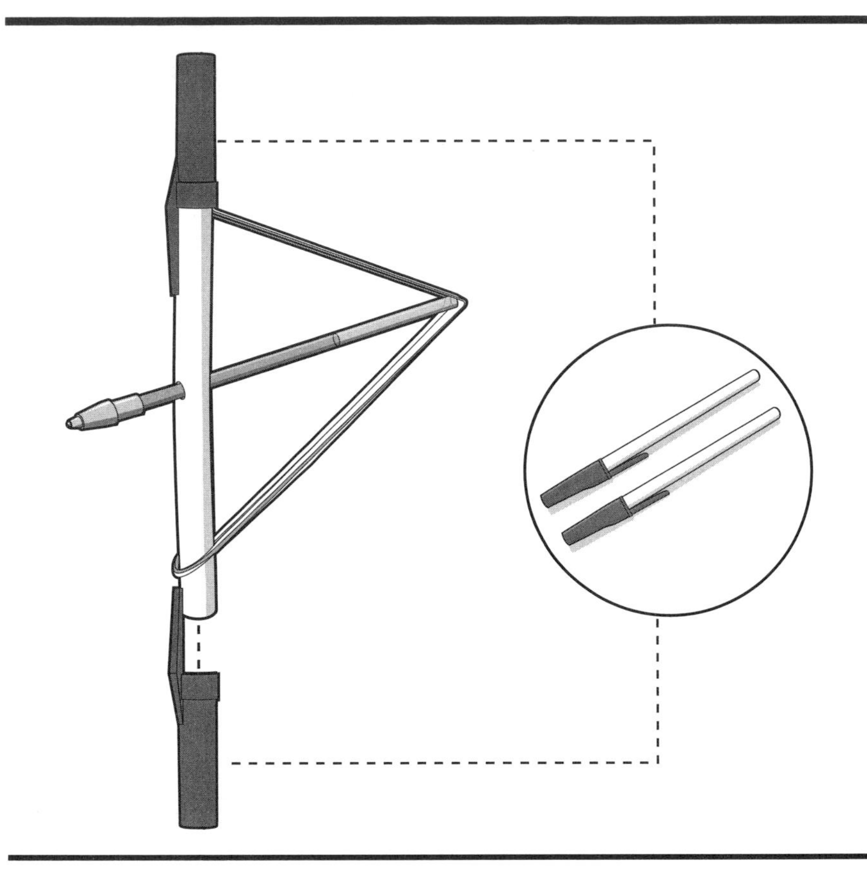

This version of the Bow-and-Arrow Pen substitutes two pen caps for the rubber band notches cut on both ends of the pen housing in the standard version. Slide the pen caps over the rubber band to secure the ends. You may add tape to help hold the caps in place. You will still need to disassemble your pen and cut a small hole in the center of the pen housing for your ink arrow.

# HANGER SLINGSHOT

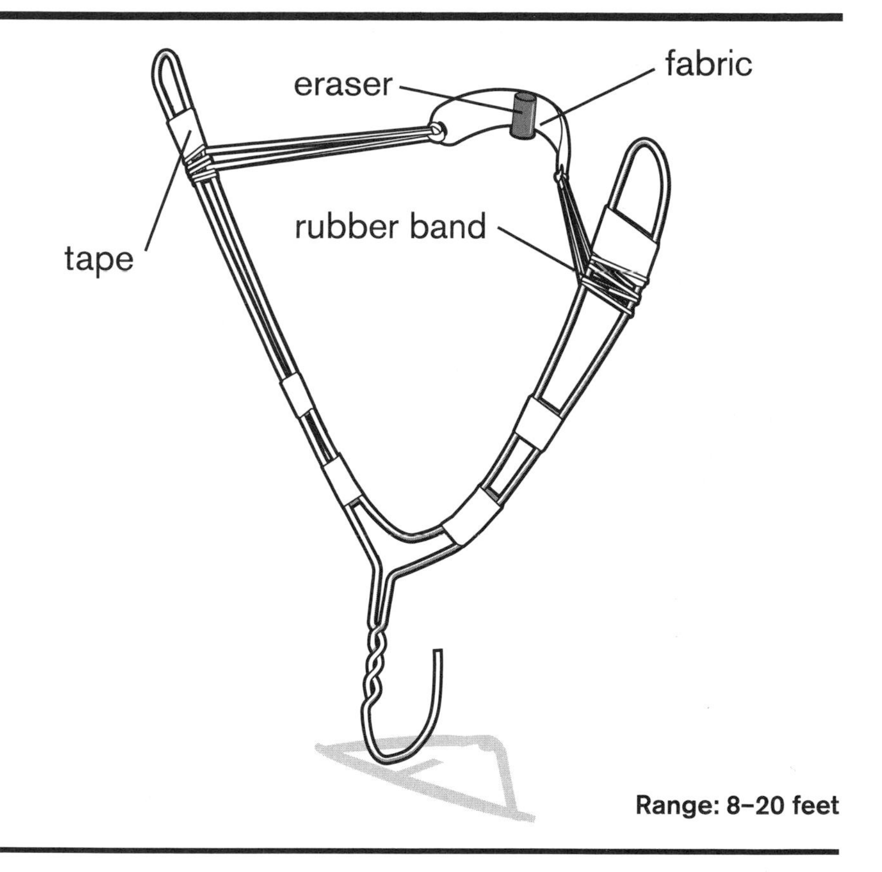


**Range: 8–20 feet**

Every urban marksman needs a trusty slingshot, and with some perseverance you'll master this handheld catapult. This homemade slingshot has a solid steel frame and two powerful elastic strips for launching projectiles at high speeds. Good luck target hunting!

## Supplies

1 metal clothes hanger
Masking or duct tape
1 piece of fabric
4 rubber bands

## Tools

Safety glasses
Scissors

## Ammo

1+ eraser

# Step 1

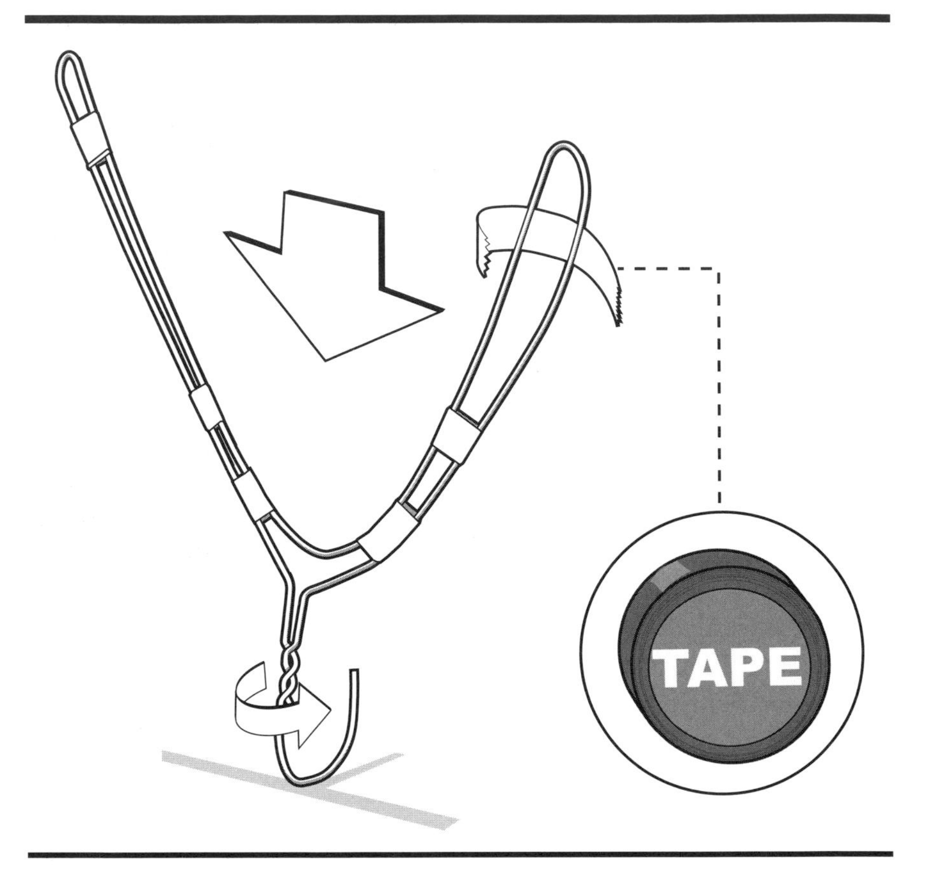


Construct the Y-frame from a simple metal coat hanger by bending the bottom bar toward the hook. Once you've shaped your frame, wrap tape around several points on the hanger to prevent the frame from bending out of shape during firing.

Also, rotate the hanger's hook 90 degrees. This will become your slingshot handle.

## Step 2

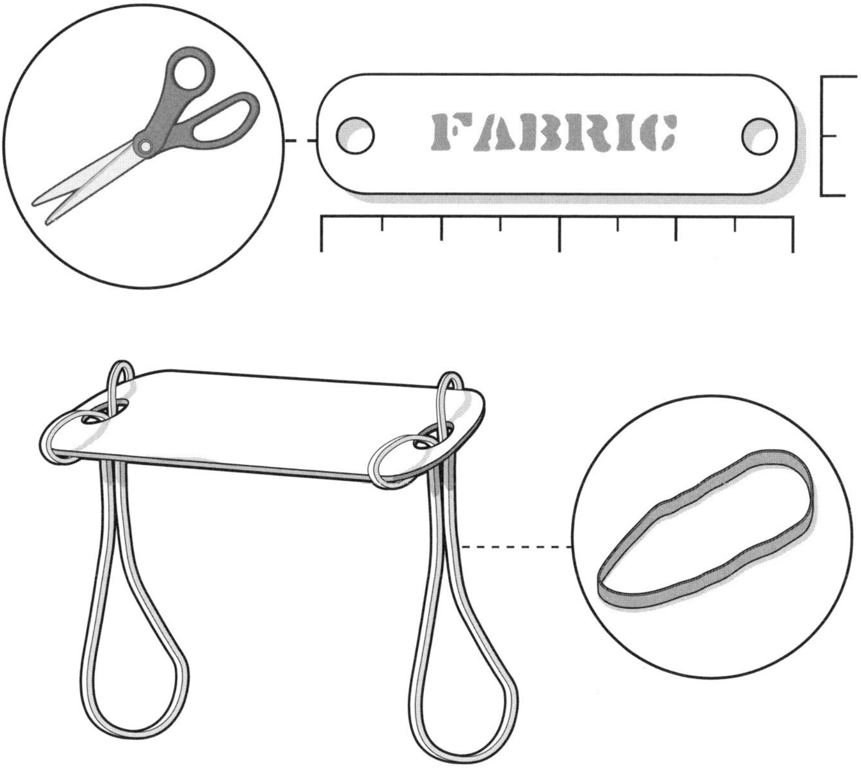

Now it is time to construct the slingshot pocket. Cut out a small piece of fabric from an old washcloth or dishrag. It should measure roughly 2 inches by 1/2 inch and have rounded corners at the ends. Use a small point to cut or poke a hole in both ends of the fabric.

Now loop a rubber band through each hole and back through itself, as shown. This will create a knot when pulled.

## Step 3

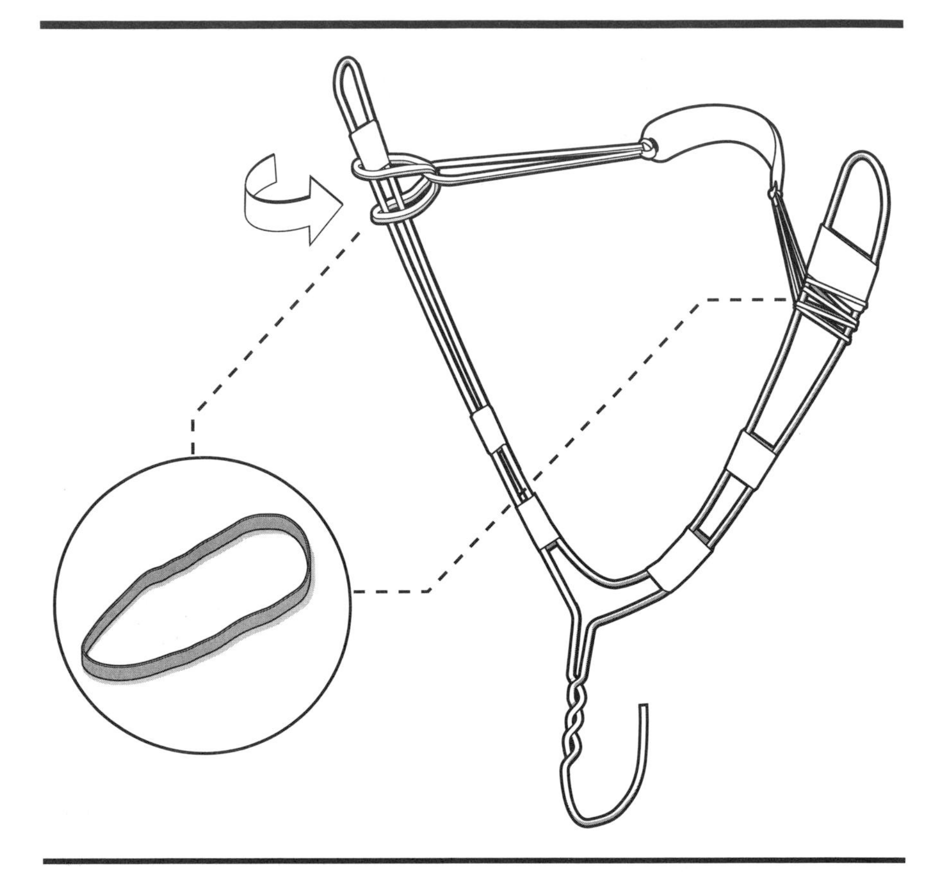

Next, use two more rubber bands to attach the power bands to the slingshot frame. Attach one rubber band per side, looping each through one of the rubber bands tied to the pouch. Then wrap each rubber band several times around its end of the metal frame. You may need to add tape or another rubber band to hold each firmly in place.

Once finished, pull back the pouch a few times to make sure everything feels right and secure. Now load your ammunition into the pouch, pull back, and release. It is important that you do not draw the power bands directly back toward your eyes when firing. If the bands become hard, brittle, or damaged, you should replace them. Remember to always wear safety glasses when shooting.

# PENCIL SLINGSHOT

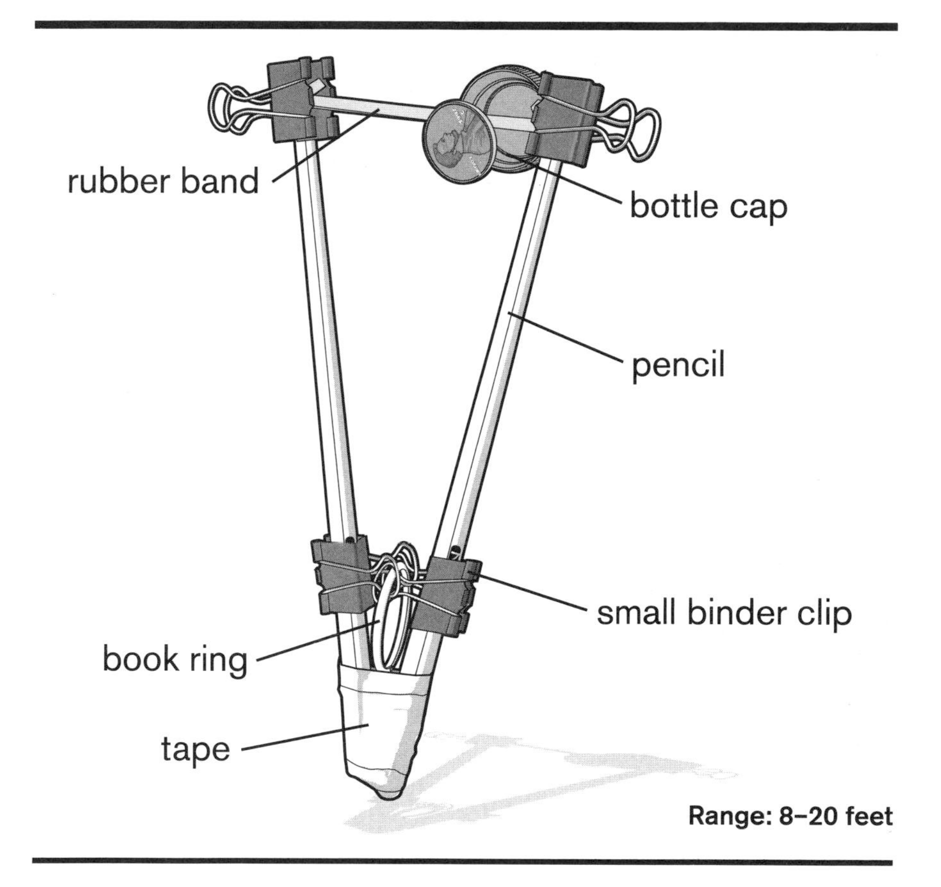


This nifty little slingshot can be quickly assembled for Monday morning target practice. Its compact design is capable of shooting coins at incredible speeds. The Pencil Slingshot is also built to last and take abuse.

## Supplies

1 plastic bottle cap
1 wide rubber band
4 small binder clips (19 mm)
2 pencils
1 book ring
Duct tape

## Tools

Safety glasses
Hobby knife

## Ammo

1+ coins

## Step 1

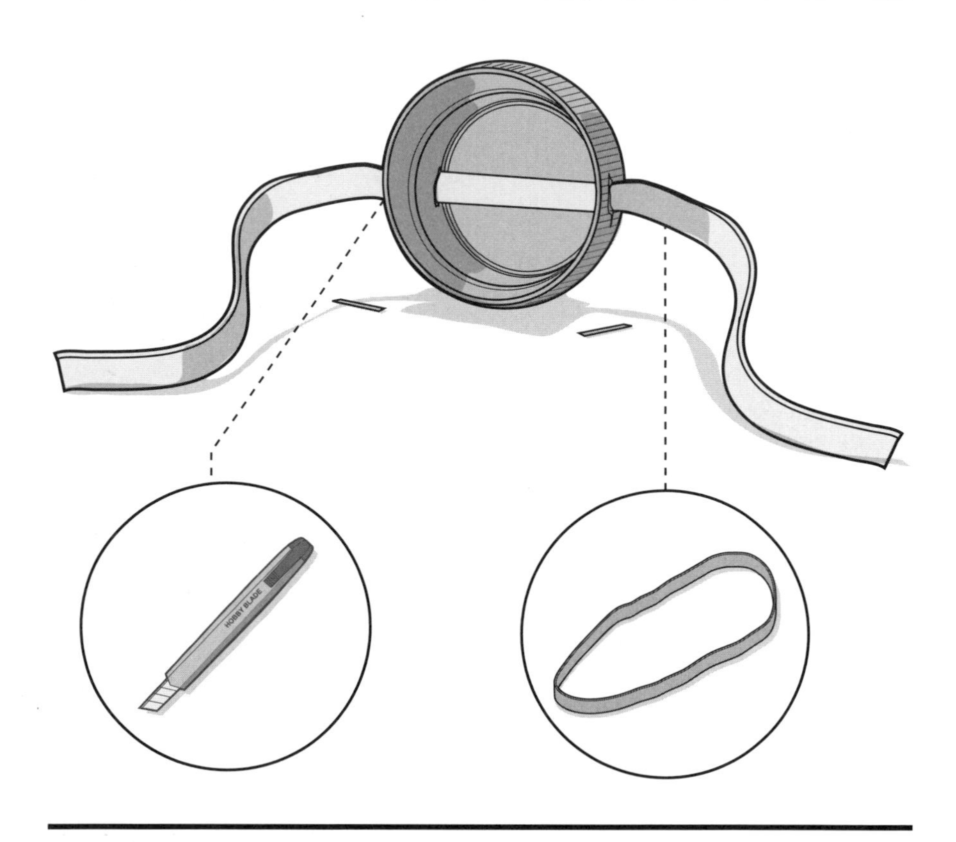


The first step is to construct the coin pouch. Use a hobby knife to cut two slits through the opposite sides of a plastic bottle cap. The slits should be approximately the same length as the width of the rubber band.

Next, cut a wide rubber band open and slide it through the slits until the cap is centered.

## Step 2

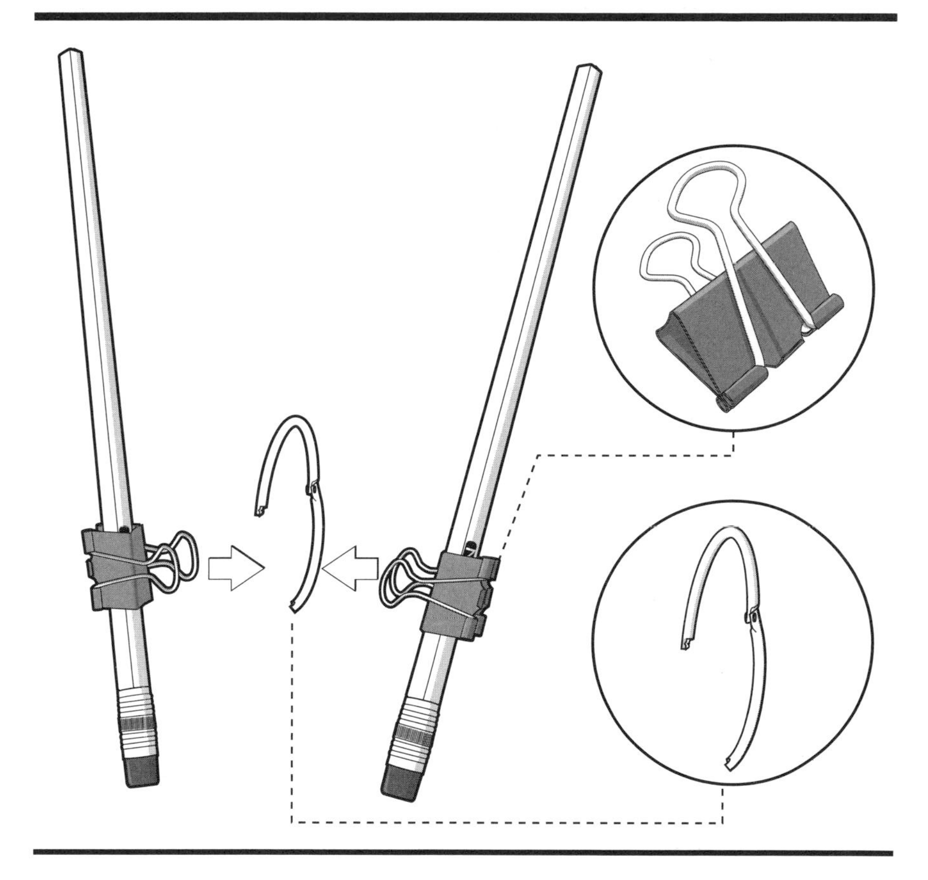

Now it's time to construct the Pencil Slingshot frame. Attach two small binder clips to two unsharpened pencils, about halfway between the bottom and middle of each.

Next, loop a metal book ring through the clips' metal handles and snap the book ring shut.

## Step 3

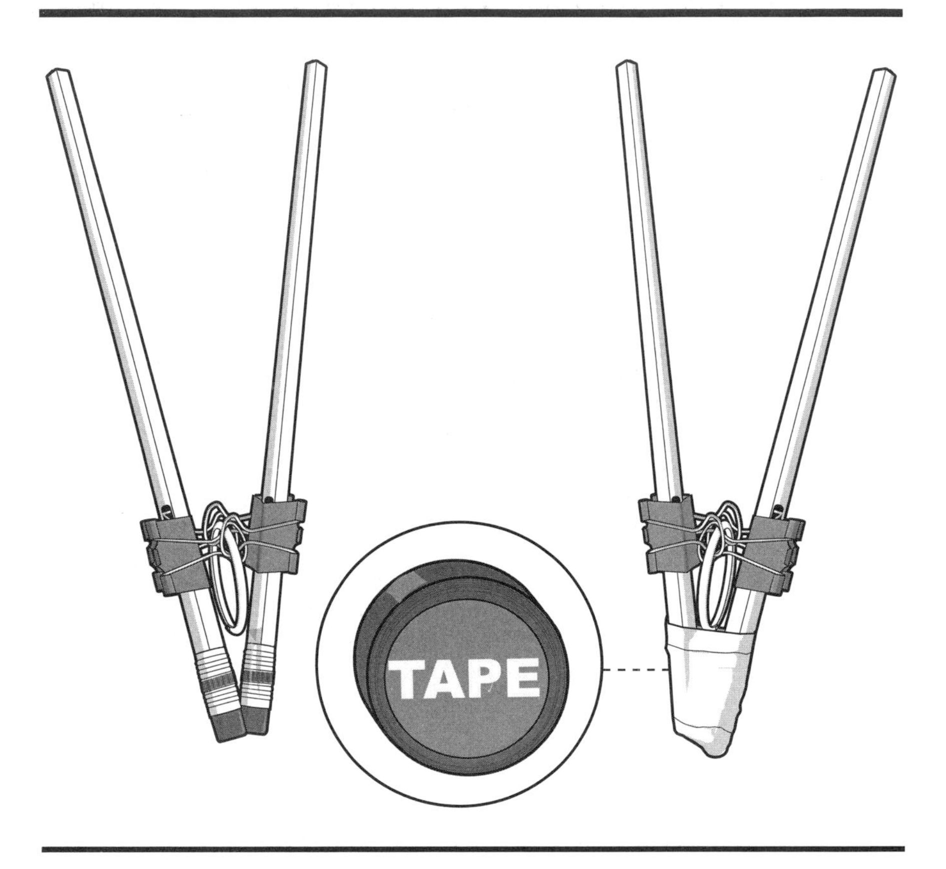


Next, secure the pencils together. Touch the two pencil erasers as shown, then tape them together. Try to prevent the pencils from overlapping and instead make sure that they are lined up; this will make for a stronger slingshot.

## Step 4

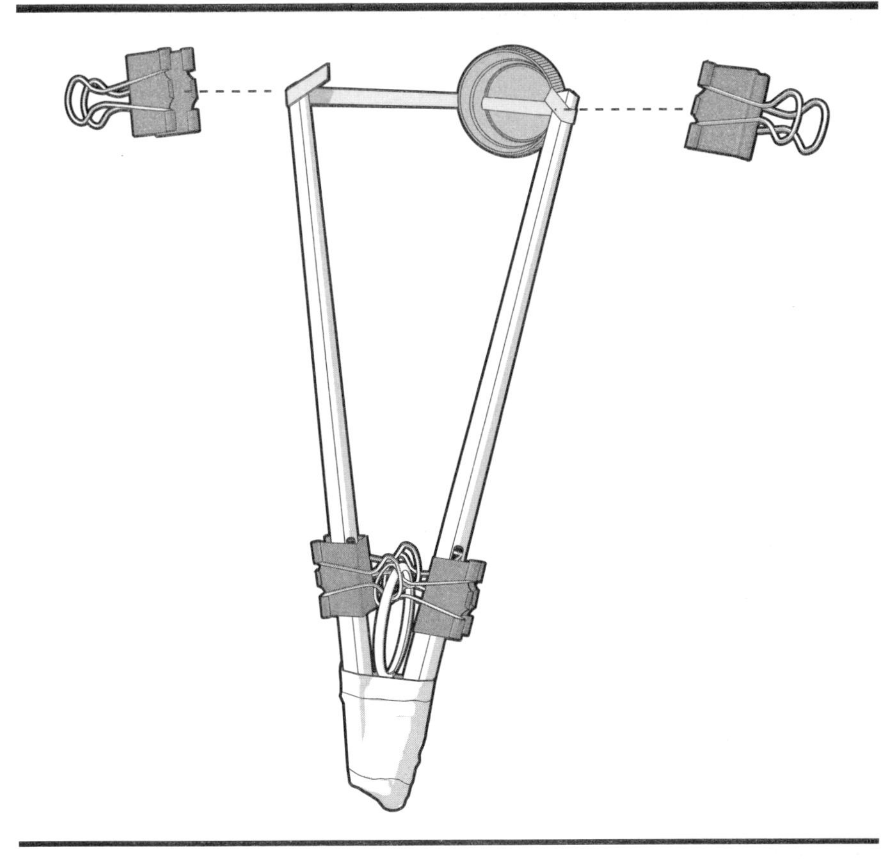

Now it's time to attach the power band to the two-pencil frame. Wrap one end of the rubber band around one pencil tip, then use a small binder clip to secure it in place. Do the same for the other side. Once assembled, pull back the power band a few times to test the strength of the band and binder clips. If your binder clips do not hold the rubber band, use additional tape to secure them.

Once everything is in working order, select a target and have fun!

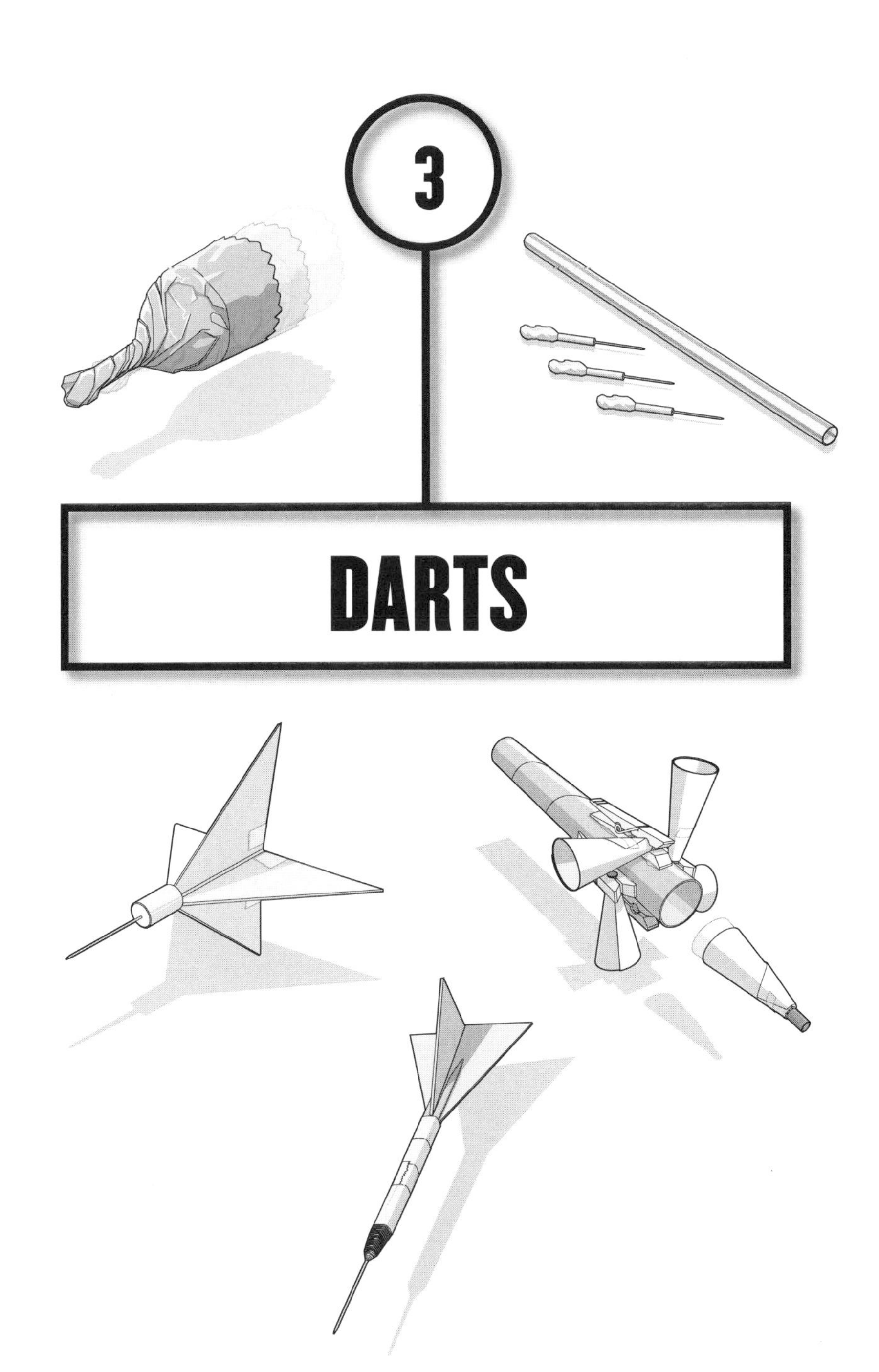

# 3

# DARTS

# BUBBLE GUM DART

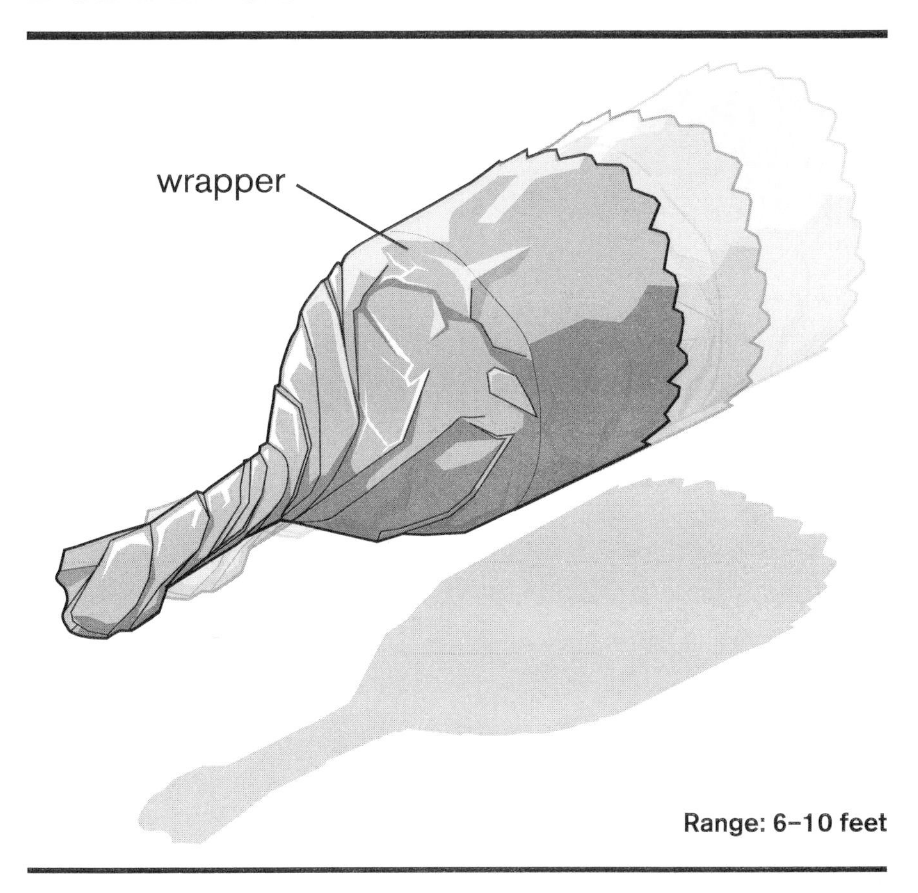


**Range: 6–10 feet**

The Bubble Gum Dart is a marvel of quick weaponry. Although it is only capable of traveling a distance of 10 feet or less, this microrocket takes mere seconds to construct, making it an absolute must for the avid chewer and skilled bubble-blowing assassin.

### Supplies

Chewing gum

### Tools

Safety glasses
Your finger
Your mouth

### Ammo

1+ gum wrapper

## Step 1

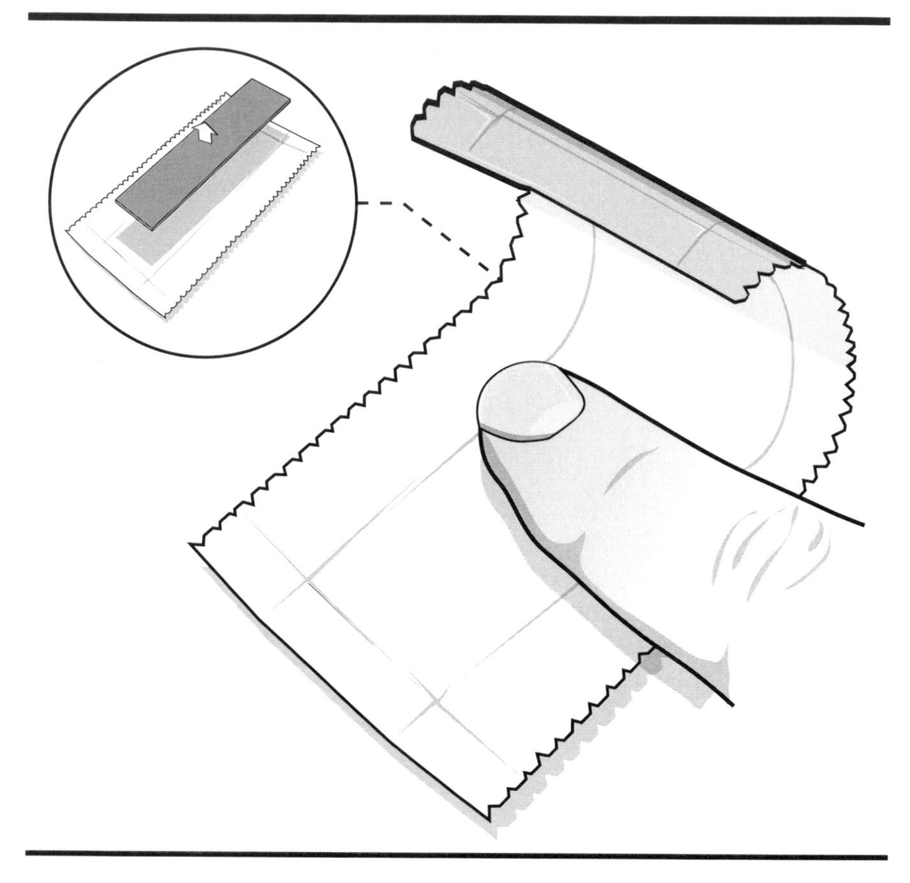

Locate an unused traditional stick of gum. Flatten out the protective foil wrapper on a smooth surface, eliminating any folds or creases. Enjoy the gum.

## Step 2

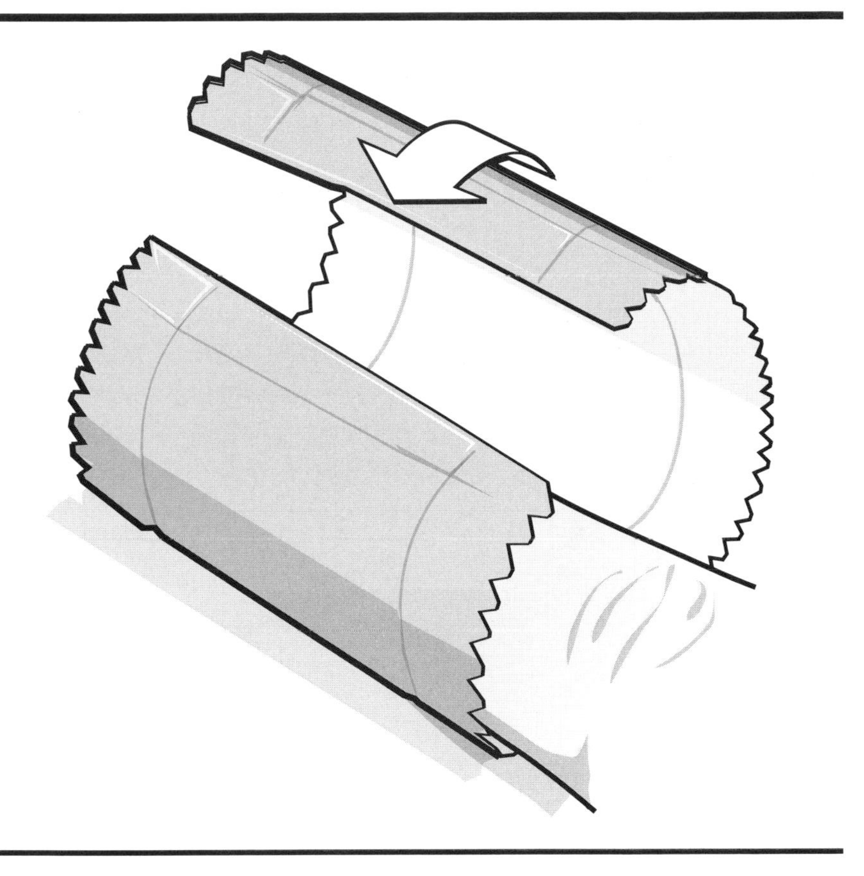

Position your finger in the center of the foil, then roll the wrapper over your finger to construct a crude cylinder. It may not be rocket science . . . but it does build a simple rocket.

## Step 3

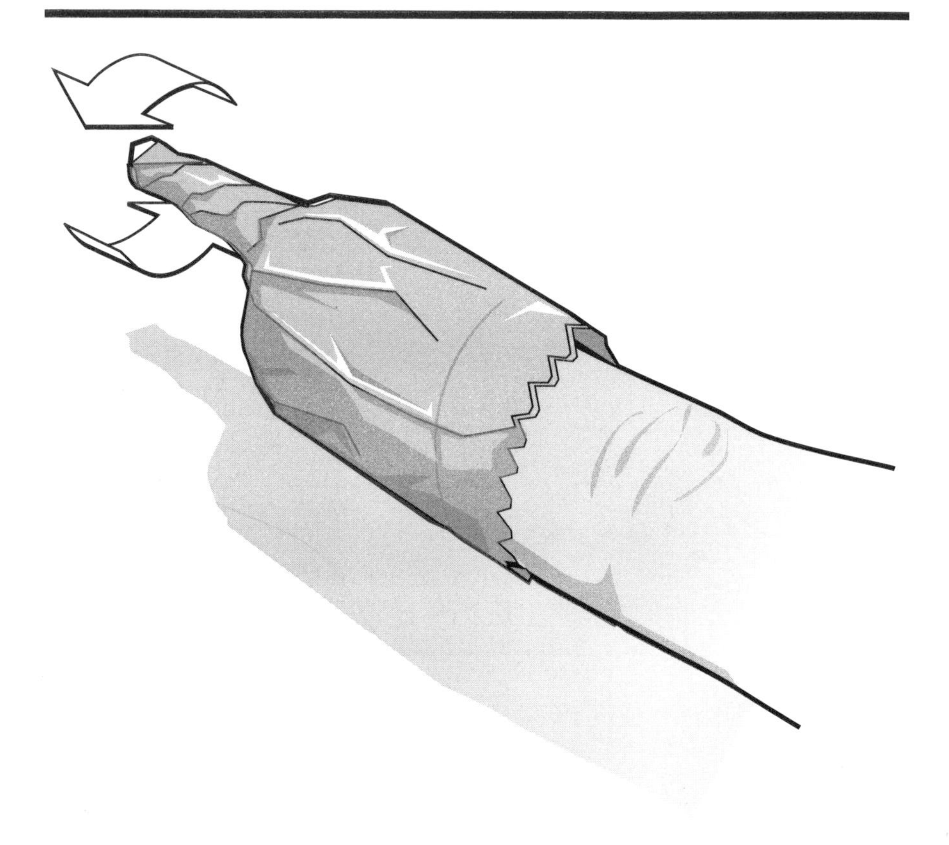

Now tightly twist the end of the wrapper in the same direction in which you wrapped it around your finger to make a crude point.

Slowly remove your finger from the cylinder. Place the Bubble Gum Dart in your mouth with only the tip protruding out between your lips, being careful not to crush the cylinder. Take a deep breath through your nose and then blow out through your mouth. Slowly open up your mouth as you exhale and watch the gum rocket fly across the room. These rockets' accuracy varies, but you'll get better with practice. Never aim at another person.

# SHOELACE DARTS

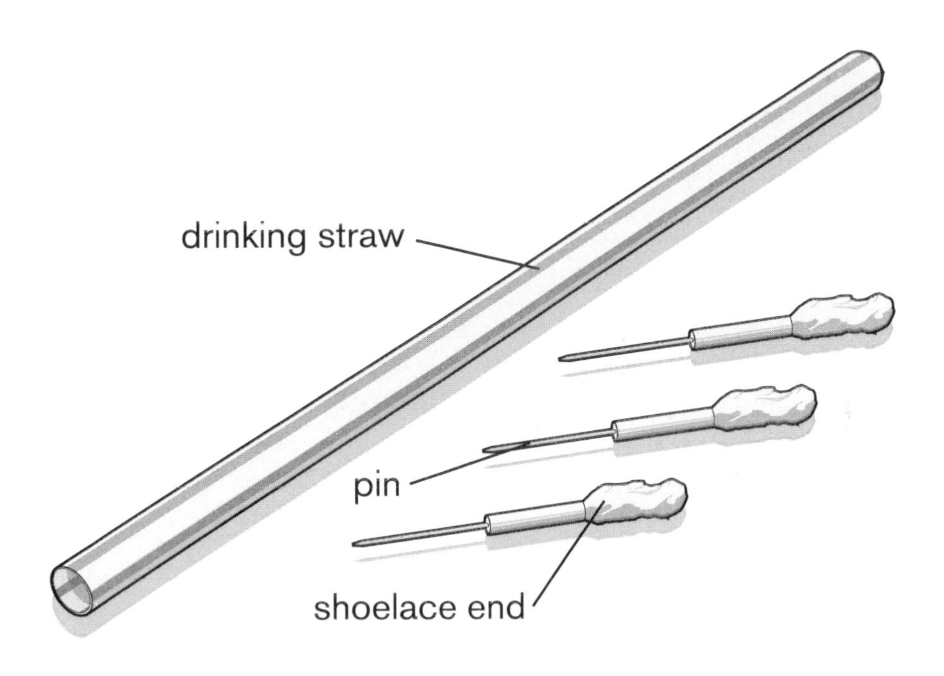


**Range: 10–20 feet**

Drinking straw to some, blowgun to others. This blowgun is capable of firing small needle-tipped darts with balloon-popping power! Unfortunately, you'll need someone to involuntarily donate their shoelace ends to create this weapon.

### Supplies

1+ shoelaces
1+ pins
1 drinking straw

### Tools

Safety glasses
Scissors
Pliers (optional)

### Ammo

The constructed darts

## Step 1

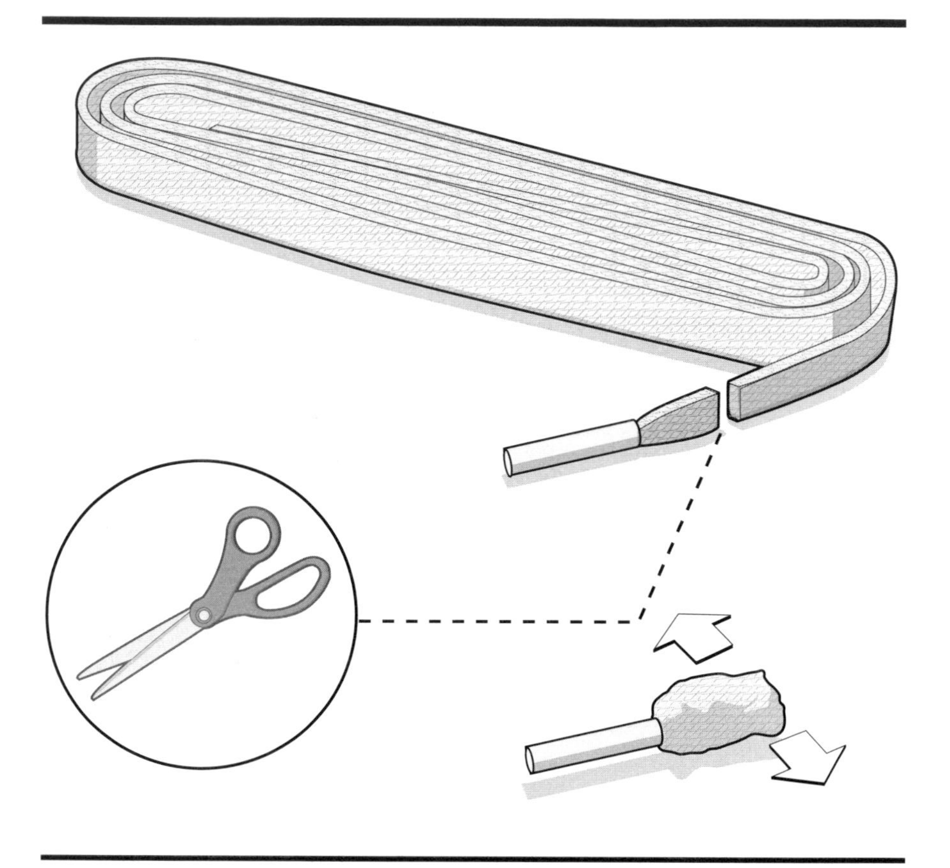

Your first plan of attack is to chop off the ends of someone's shoelaces. Take each detached end and use your fingers to fluff out the shoelace fabric while keeping together the plastic-tipped cylinder (also know as the aglet). This fluffed-out fabric will become your tail section and will help control the trajectory.

## Step 2

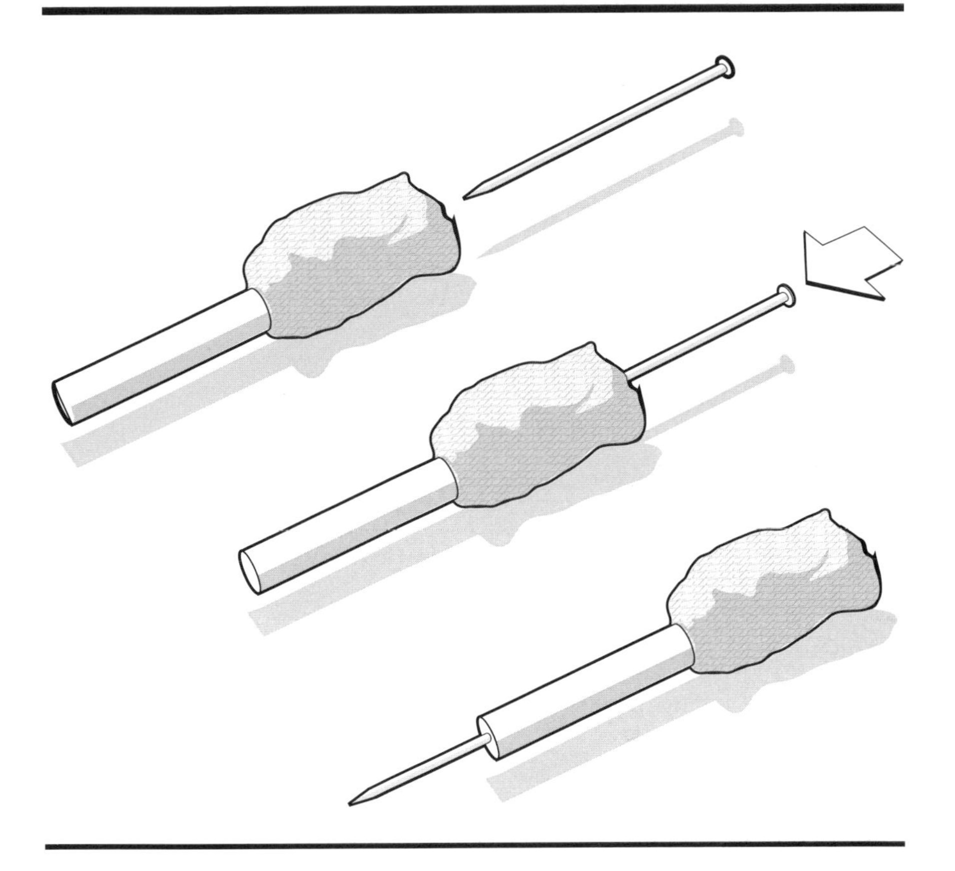

Now you must equip your dart with a point—a straight pin. Depending on how the shoelace was manufactured, you may run into a problem on this step; however, there are ways around it. The ideal method is to push the pin through the back tail section so that the point protrudes out of the end. If you find this difficult, you may want to use pliers to hold onto the lace to avoid poking your skin.

If the plastic aglet is too tight, change your approach: first poke a hole in the front end of the plastic aglet, then stick your pin in backward. You will avoid having to burrow all the way through the tip, but the pin will be less secure.

Whatever method you decide to execute, your final product should be a tightly secured pin sticking out of the aglet end.

## Step 3

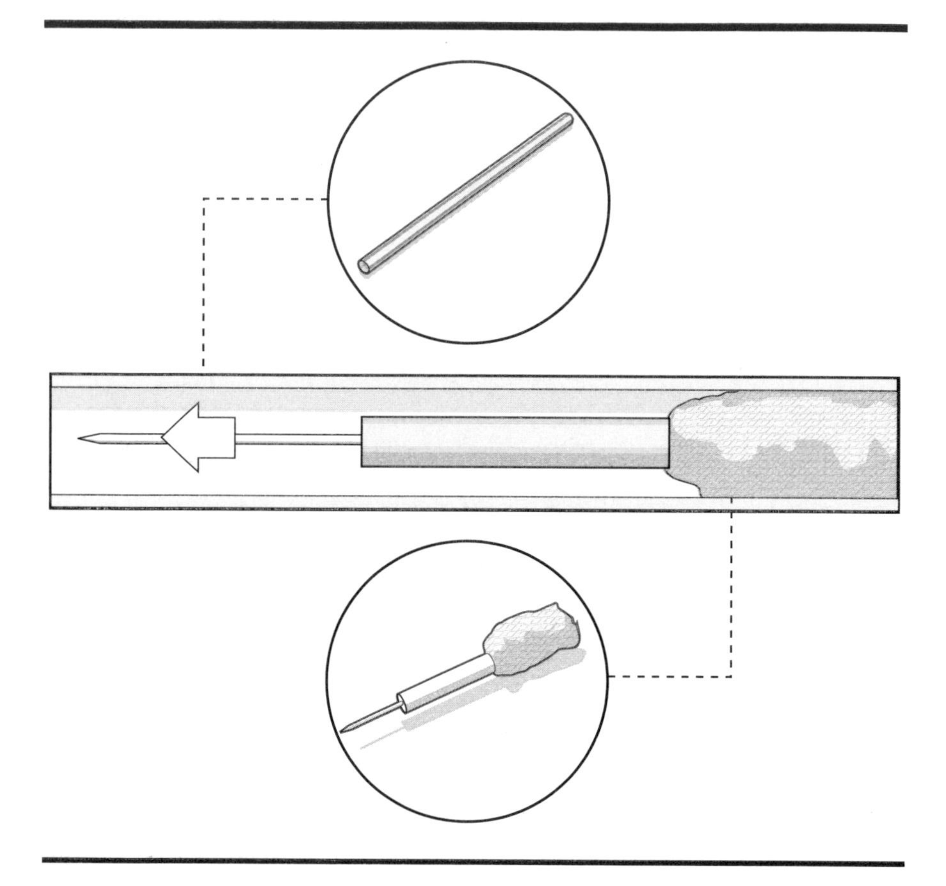

Now stuff your dart into a drinking straw with the point facing toward the exit. Aim the straw and hold it to your lips. Then take a deep breath and exhale sharply to blow the dart out.

It is very important to remember that you're firing needles from your mouth. ***You should never inhale while the straw is in your mouth waiting to be launched.*** Always be responsible and fire it in a controlled manor. ***Safety glasses are a must, as is staying clear of spectators.***

# ERASER DART

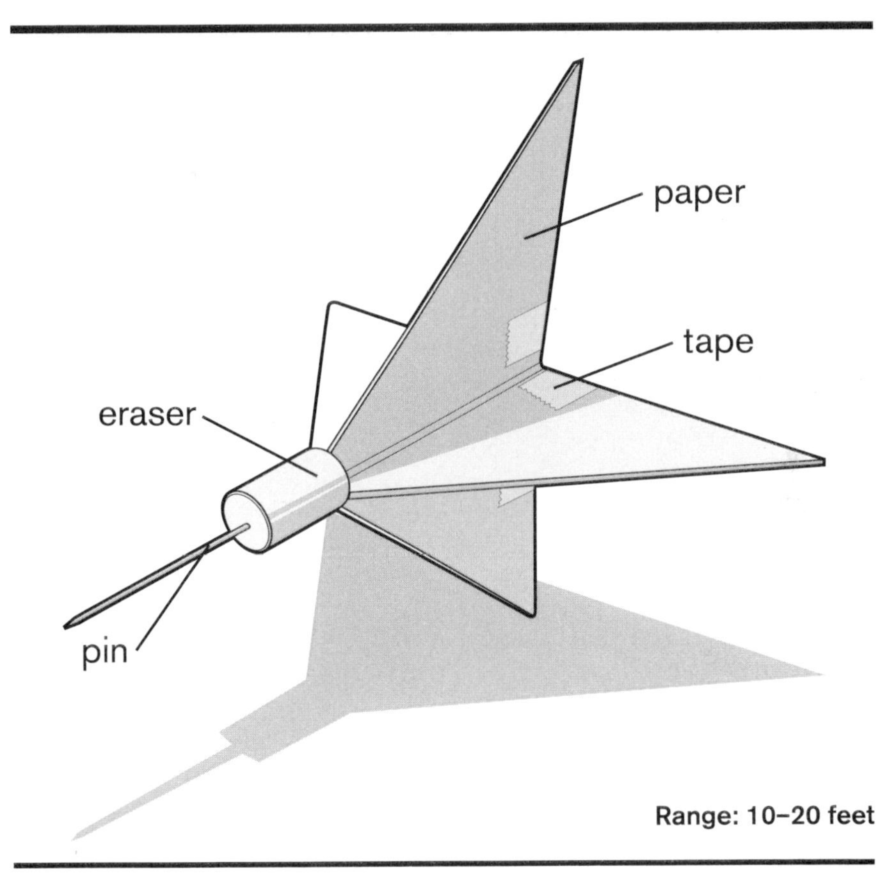


**Range: 10–20 feet**

The Eraser Dart is the official dart of the little-known National Eraser Dart Association (NEDA). Because NEDA is so secretive, you're encouraged to make up your own rules, and with such loose standards, starting a NEDA league is simple and fun. The only requirements for a friendly tournament are to always throw at a makeshift dartboard, not at one another, and to keep all players behind a throwing line marked on the floor.

### Supplies

1 sheet of paper
1 pencil eraser
1 pin
Transparent tape

### Tools

Safety glasses
Scissors
Pocketknife

### Ammo

The constructed dart

## Step 1

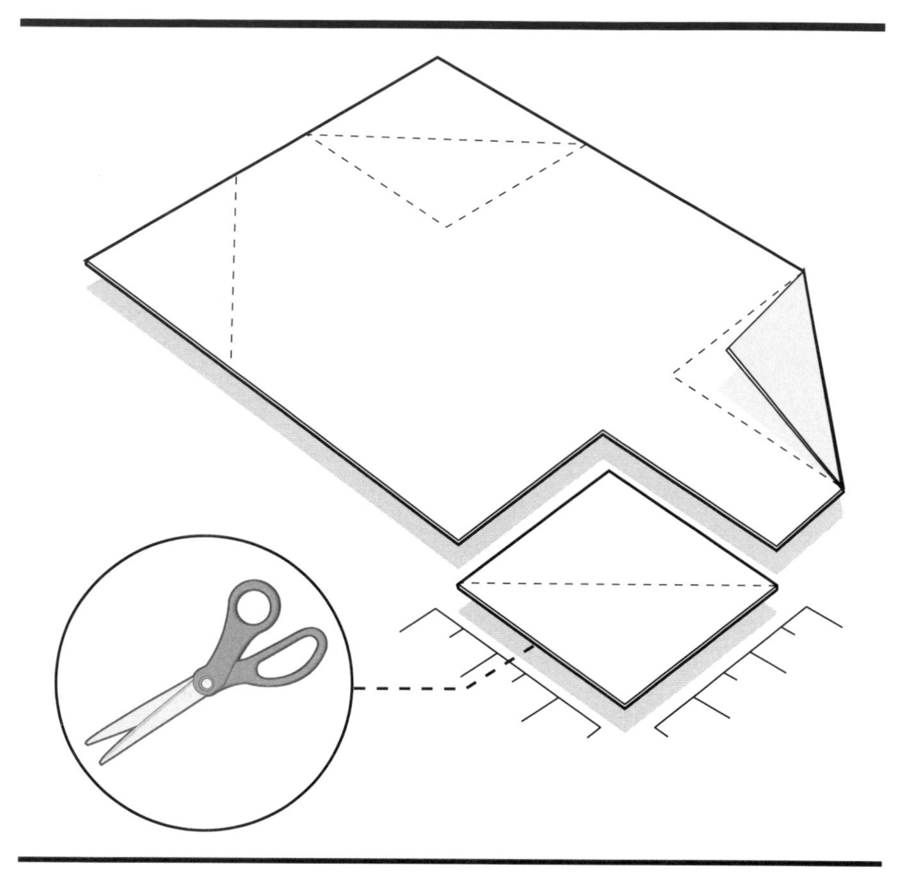

First, construct the tail of the Eraser Dart. Using a normal-sized sheet of paper, fold up one of the corners to be approximately 2½ inches by 2½ inches. This will form a square once it's unfolded.

After you have made a perfect folded square, cut it out. This is all the paper you will need to make one Eraser Dart; save the rest of the paper to make additional darts later.

## Step 2

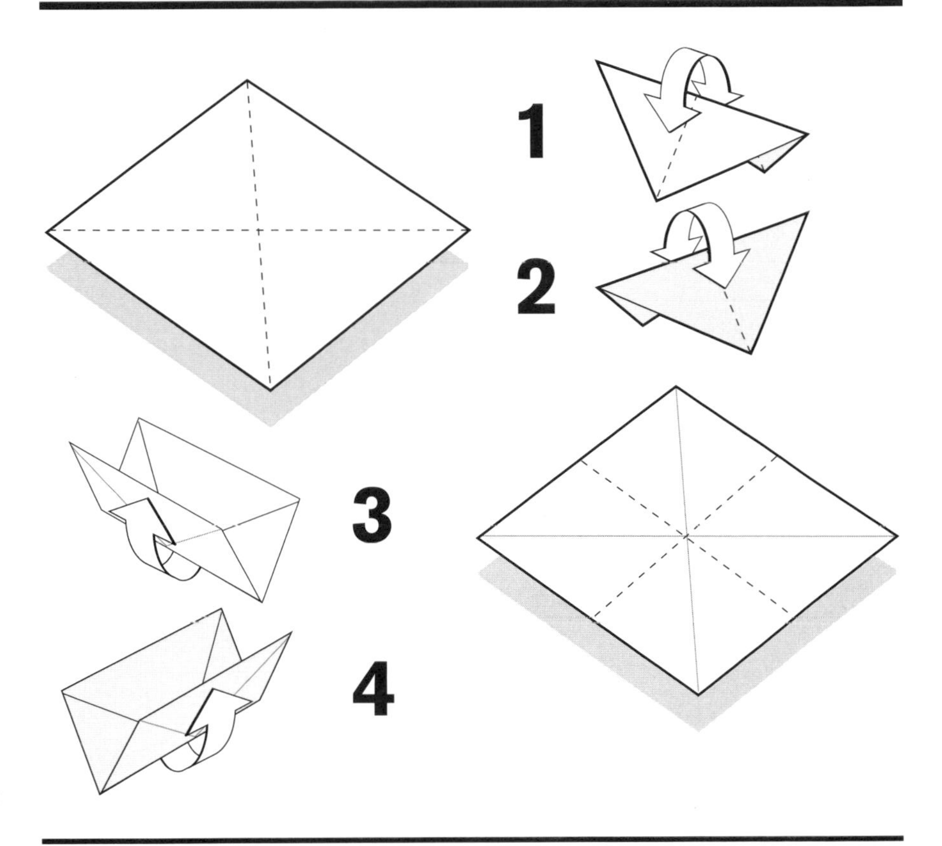


Now that you have a perfect square, you will need to add some additional folds. This step is similar to the Water Bomb assembly found in chapter 6 (page 207). First, fold opposite corners of the paper together to form two triangles, then unfold the paper and fold the other two corners together to form two triangles, as shown above.

Next, fold the paper in half both ways. Once you have finished you should have folded the paper a total of four times and should have four crease lines as shown above. These creases will act as guides for the major folds in the next step.

## Step 3

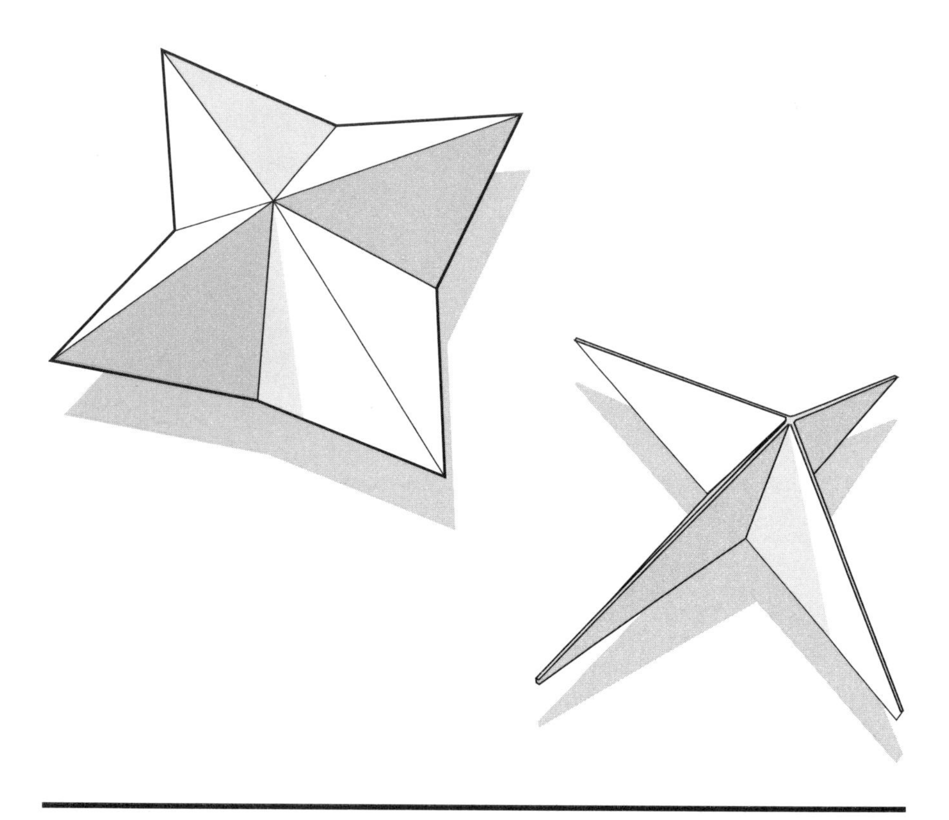

Now, creatively folding your square, push up on the center point and down on the middles of the sides until you have what appears to be a star. This may take a few attempts due to the confusing nature of the creases. Use the illustration above for reference.

Once you've completed the folds, run your fingers along the fold lines to crease the edges. This will help the paper keep its form during construction.

## Step 4

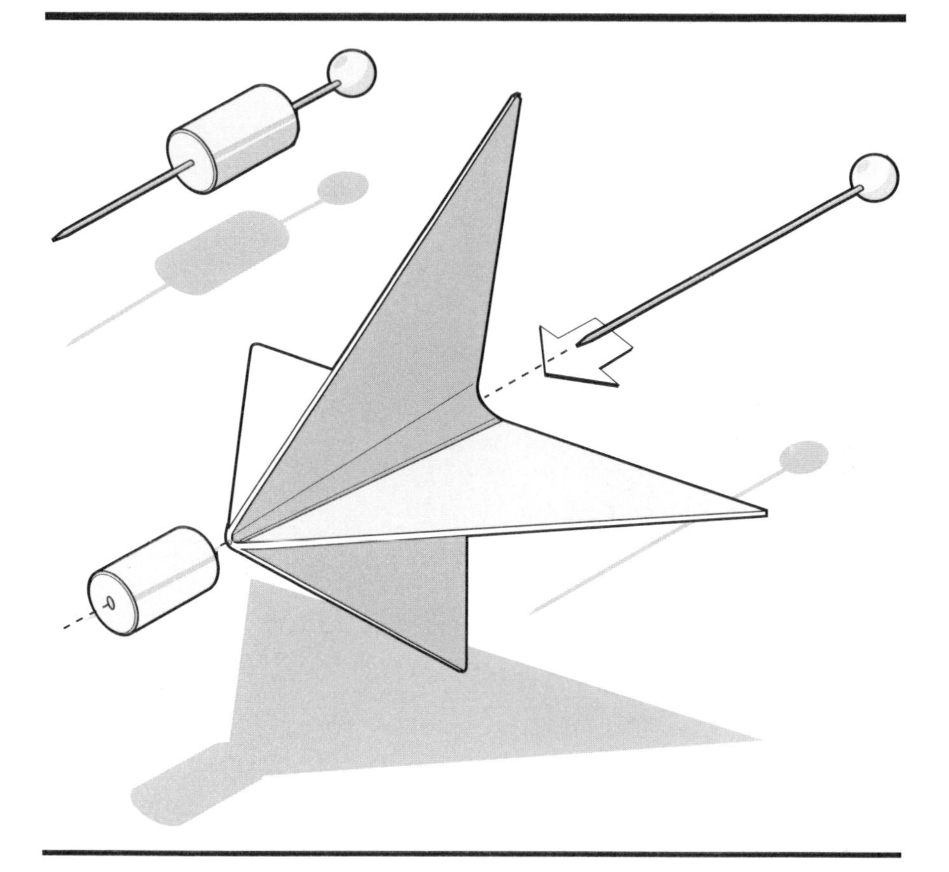

Cut a new eraser off a pencil using a pocketknife. Prepare the eraser by inserting a pin through the center and then removing it.

Now open up the rear section of the paper fins. From the back, push the pointed end of the pin through the center point of the paper fins. (A standard straight pin with a round head is ideal for this dart design. A smaller pinhead will ultimately fall out of the front of the paper fins and cause failure.) Once through the paper, push the pin back through the hole you made in the center of your eraser. Make sure the head of the pin is tight against the eraser for a solid assembly.

## Step 5

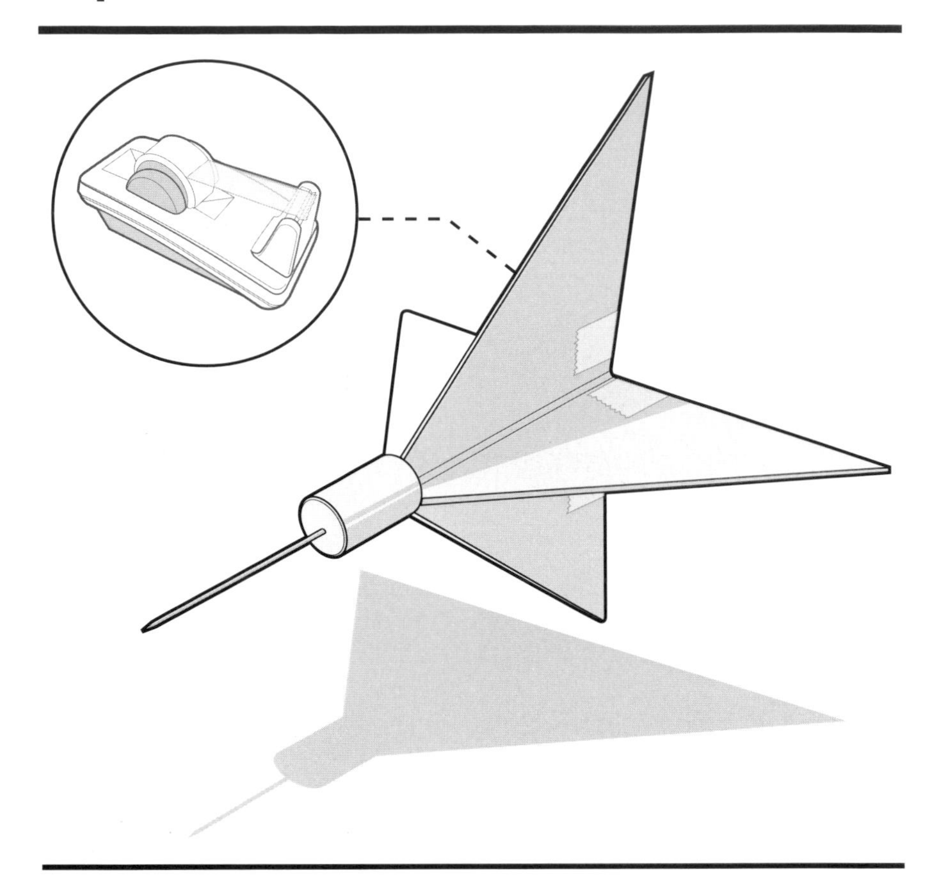

Your dart is almost complete—just tape the backs of the wings together so that the Eraser Dart keeps its shape when flying across the room.

Safety first! Your minijavelin should travel with enough force to stick into a cork or wooden target, but these Eraser Darts are not toys and should never be thrown at human or animal targets. It is important that you aim at a large target area away from spectators when launching your darts. Remember, because these darts are homemade, accuracy will vary.

# LONG DART

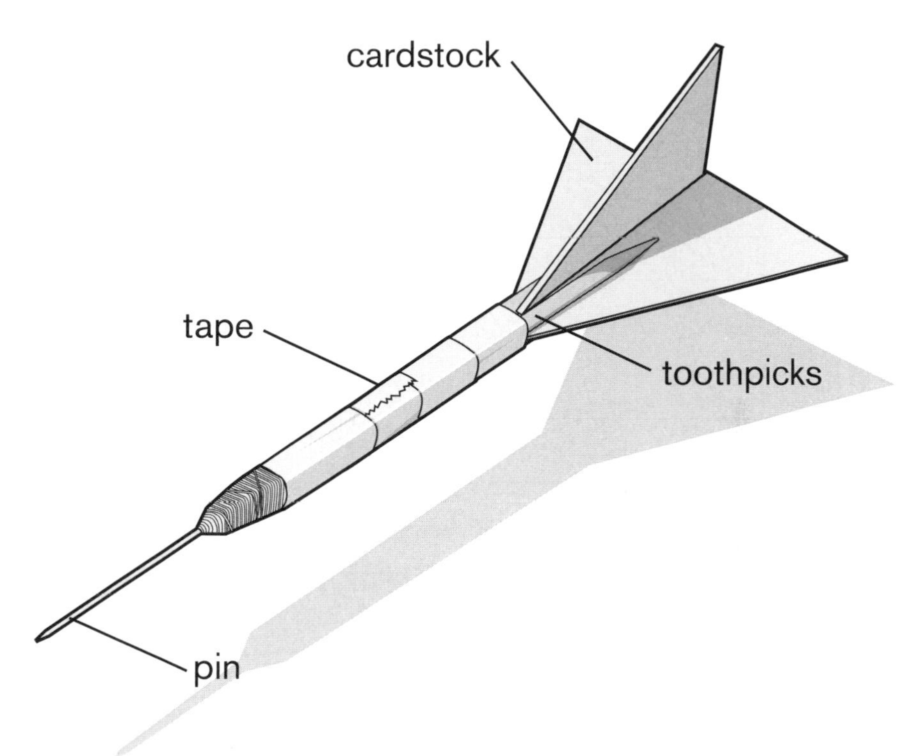


**Range: 10–20 feet**

Dartboards beware! Long Darts are designed for speed, distance, and accuracy. These microarrows are perfect for competitive activities, such as traditional pub darts. Because they cost only a few cents to build, accumulating a bunch of them for friendly homemade games of Round the Clock or Jumpers is no problem at all. Don't have a dartboard? Chapter 7 (page 240) has a perfect one for you to use for game after game.

### Supplies

4 toothpicks
Masking tape
1 small metal pin
Thread
Glue (optional)
Cardboard

### Tools

Safety glasses
Scissors

### Ammo

The constructed dart

# Step 1

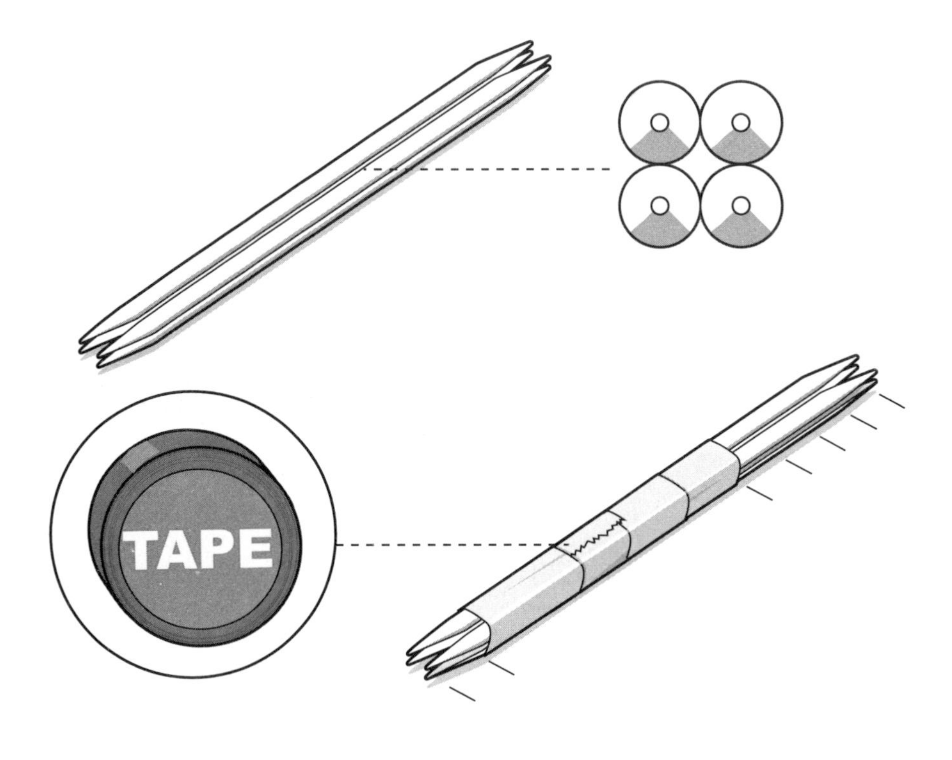


Tightly tape together four round toothpicks to form a square bundle as shown. It is important to leave the back half of the bundle untaped so you can later slide the cardboard fins into place.

## Step 2

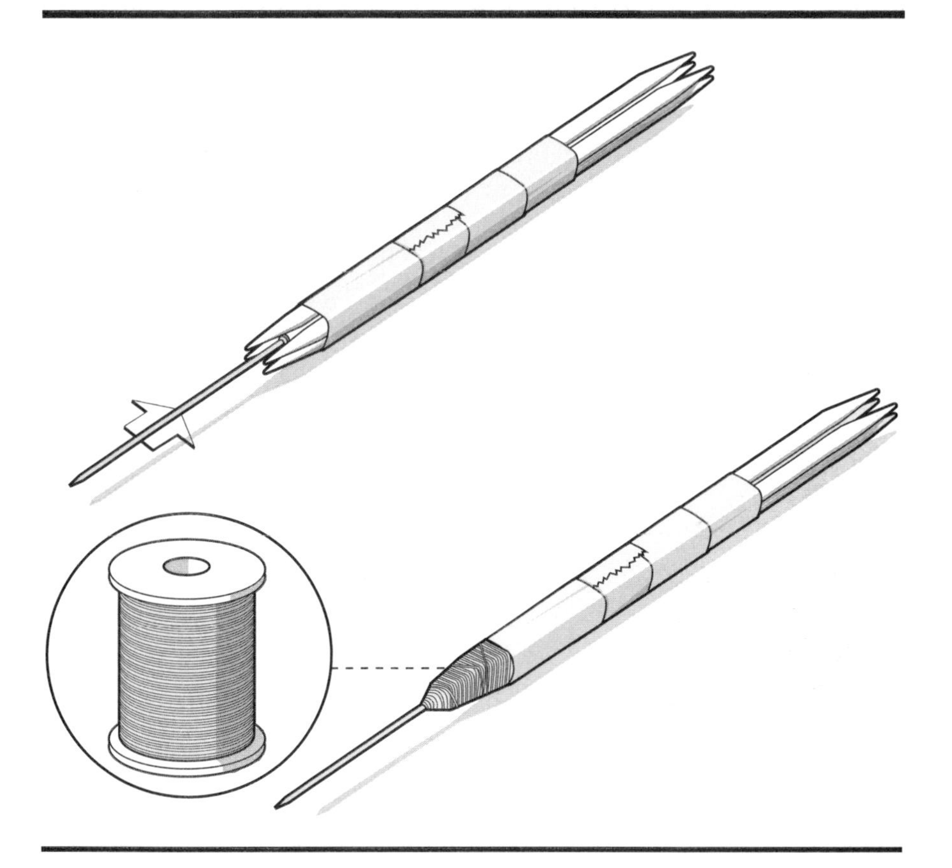

Now take a small metal pin and wedge it into the front center of the four toothpicks. A pin with a small head or a small finishing nail works best. Once in place, tightly wrap the front of the dart with thread. Continue to wrap it until your point is immovable.

You may dab some glue onto the thread wrapping to bond it further. Allow 30 minutes for the glue to dry.

## Step 3

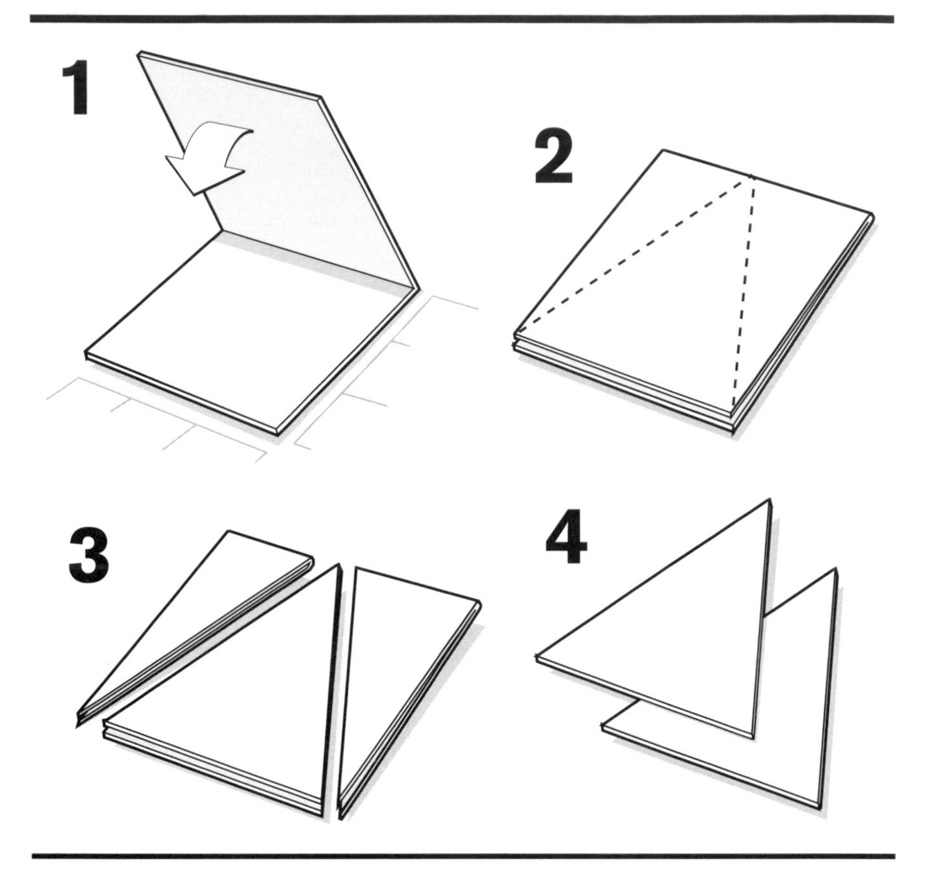


Now it's time to construct the Long Dart's fins. Use thin cardboard of the sort found on most cereal and cracker boxes. Once you have removed a section of cardboard from the box, cut out a 3-inch-by-1½-inch rectangle. Then fold that rectangle in half to create a 1½-inch-by-1½-inch double square. The square will be two cardboard layers thick. This will ensure that the four fins are the exact same size after you cut them out.

Now use scissors to cut out a triangle shape, as shown above, from your folded cardboard. Remove the extra material from both sides. When finished, you should end up with two triangles of the exact same size. The triangles should be separate, not connected at the tip.

## Step 4

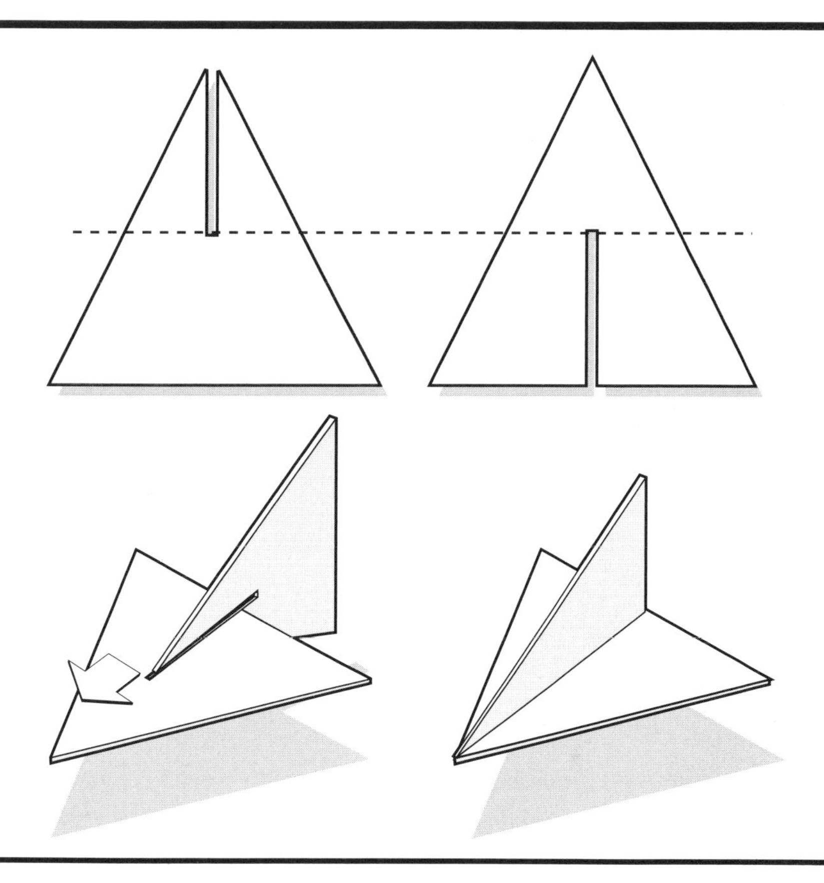

Place the two separate cardboard triangles side by side.

On the first triangle, use scissors to cut out a small slit from the top point of the triangle to about halfway down. The width of the slit should be the same as the thickness of the cardboard, but not bigger. On the second triangle, cut out a small slit of the same width from the midpoint of the bottom edge to approximately halfway up the triangle.

Now slide the two triangles together as shown to form the rear fin assembly.

# Step 5

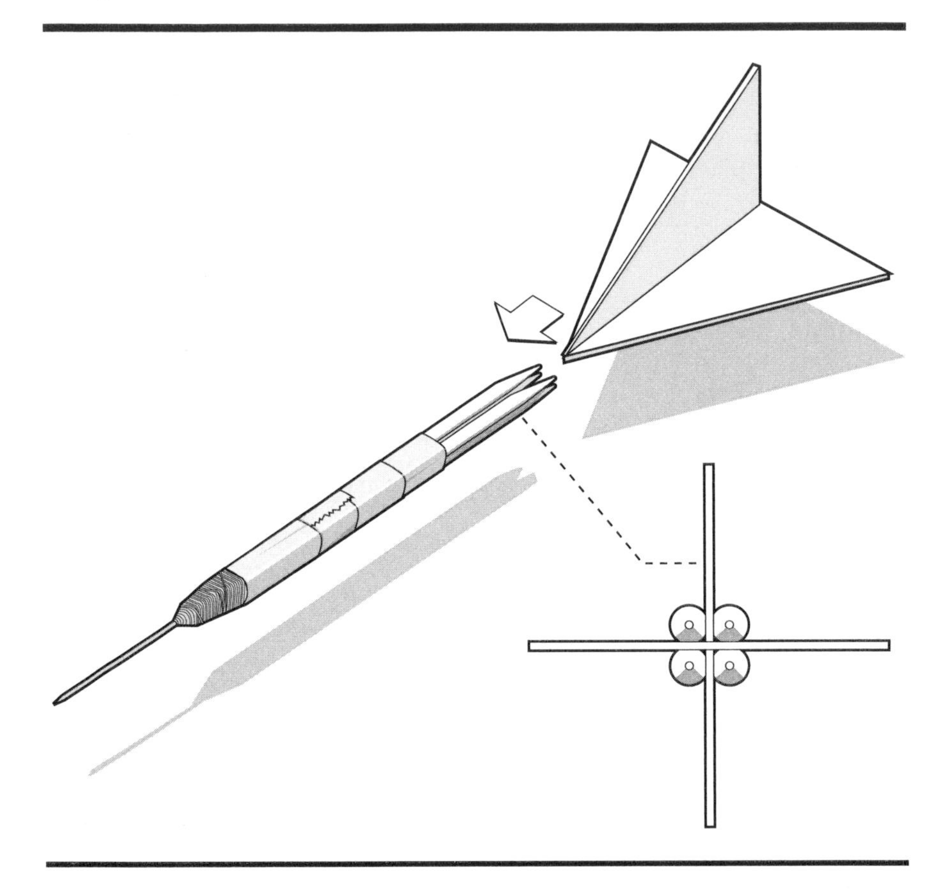

Now it's time to introduce the two parts you have created to one another. Slide the fins into the four-toothpick assembly from behind, so that one fin is wedged between each toothpick pair. The pressure of the toothpicks will hold the fins in place when in use. The Long Dart is now complete and ready for bull's-eye practice.

It is important to remember that your dart has a dangerous point at the end and is not meant for living targets. Malfunctions do occur, so use the utmost caution when launching your homemade darts. A dartboard layout, perfect for target practice, is located in chapter 7. Always use common sense, and use at your own risk.

# PAPER DARTS

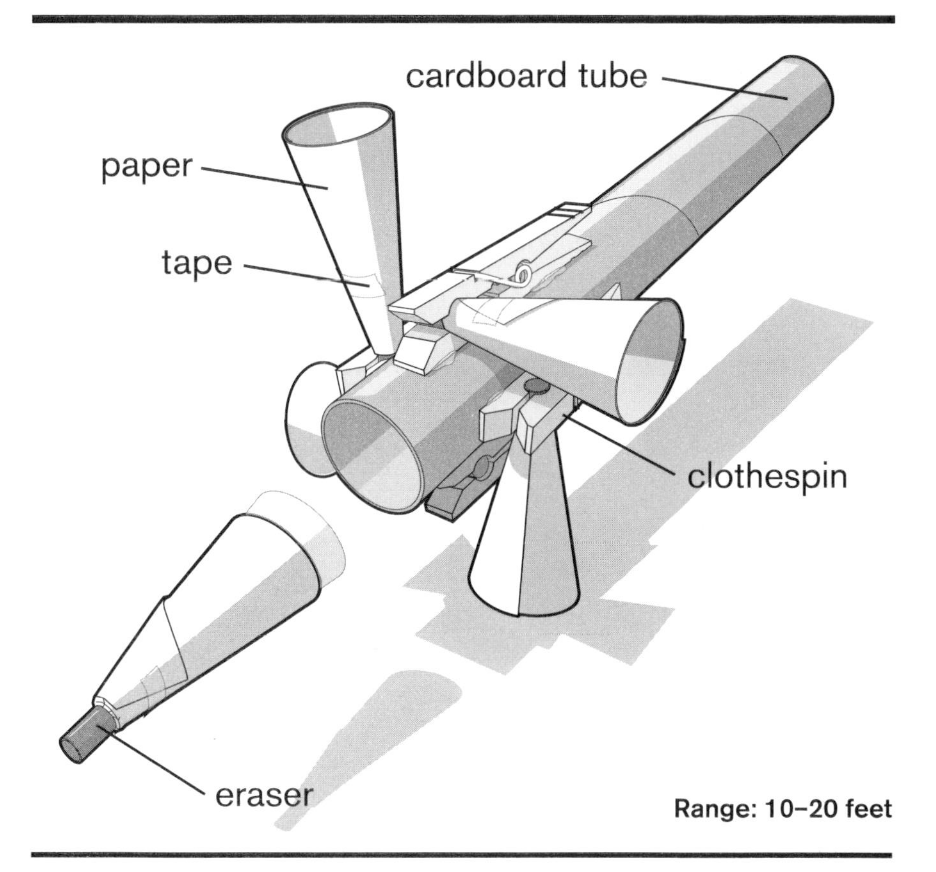


**Range: 10–20 feet**

The Paper Dart is the perfect indoor launcher for a friendly game of tag. Armed with several darts, you can quickly reload and refire while on the offensive. Its quick construction and harmless darts make this Mini-Weapon a must for any rainy day! Plus, once you learn the basic fundamentals of the blowgun, you can easily adapt it to other designs.

## Supplies

1 sheet of paper
Transparent tape
1 cardboard tube
4+ pushpins
4+ pencil erasers
4 clothespins

## Tools

Safety glasses (if playing Paper-Dart Tag)
Scissors
Pocketknife
Hot glue gun

## Ammo

The constructed darts

## Step 1

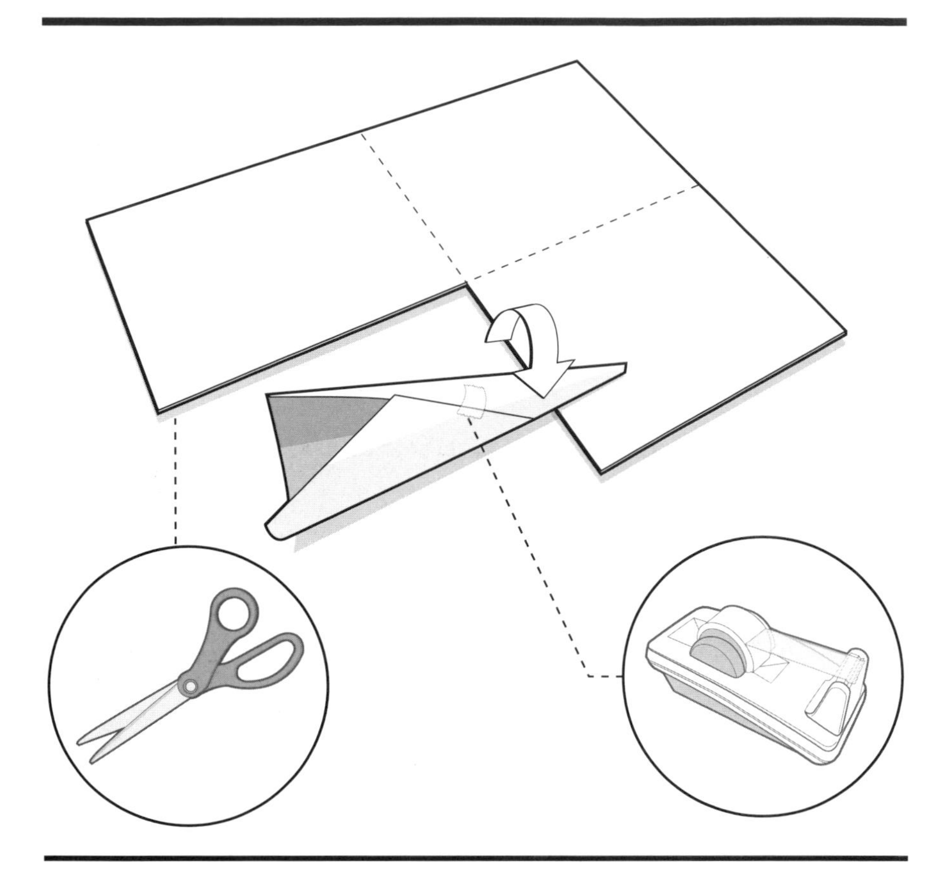

Take a standard sheet of paper and divide it into four equal sections by folding it twice. Once you have created your fold lines, cut out the sections with scissors. Next, take one paper section and roll it into a funnel. Use tape to hold the paper in place, and then make three more funnels with the remaining paper.

# Step 2

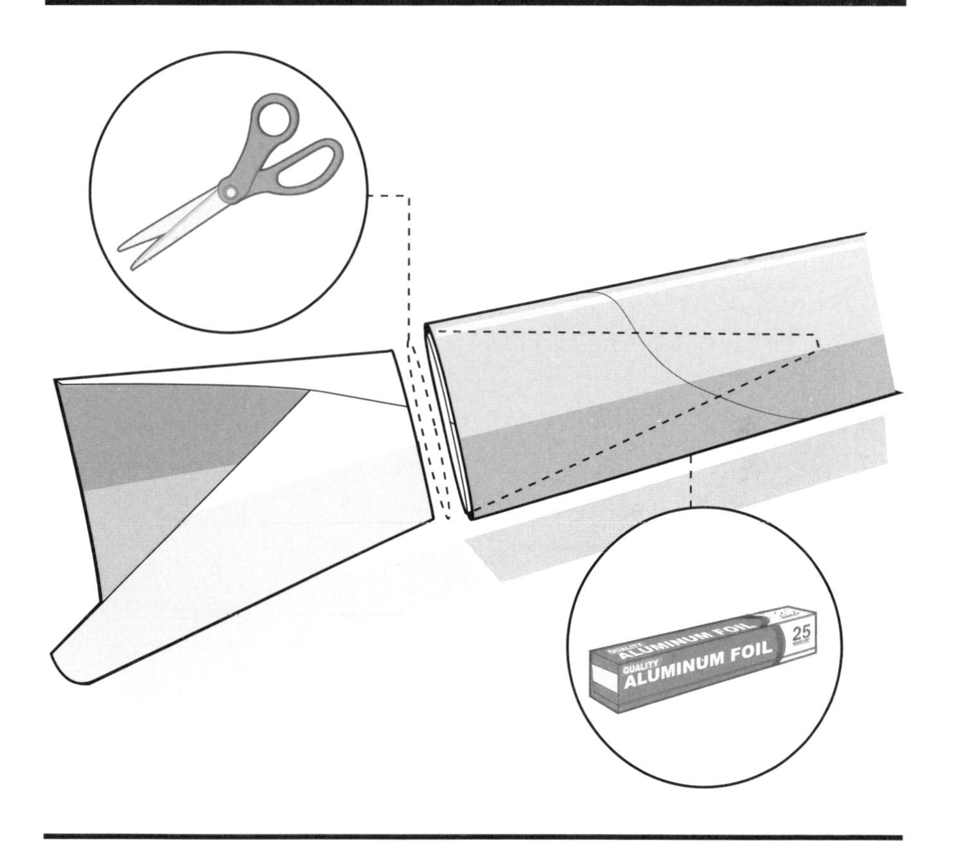


Select a smaller cardboard tube to construct your blowgun housing. An aluminum foil or plastic-wrap tube from the kitchen will work best; avoid wider tubes since they will dramatically decrease the firing distance of your Paper Dart.

Stick one of the paper funnels into the tube, being careful not to damage it. Once it's gently wedged into the tube, use scissors to trim off the extra material hanging out of the tube. Now your dart's maximum width is the exact size of the tube's diameter, which will increase the contact surface area when you blow through the tube.

## Step 3

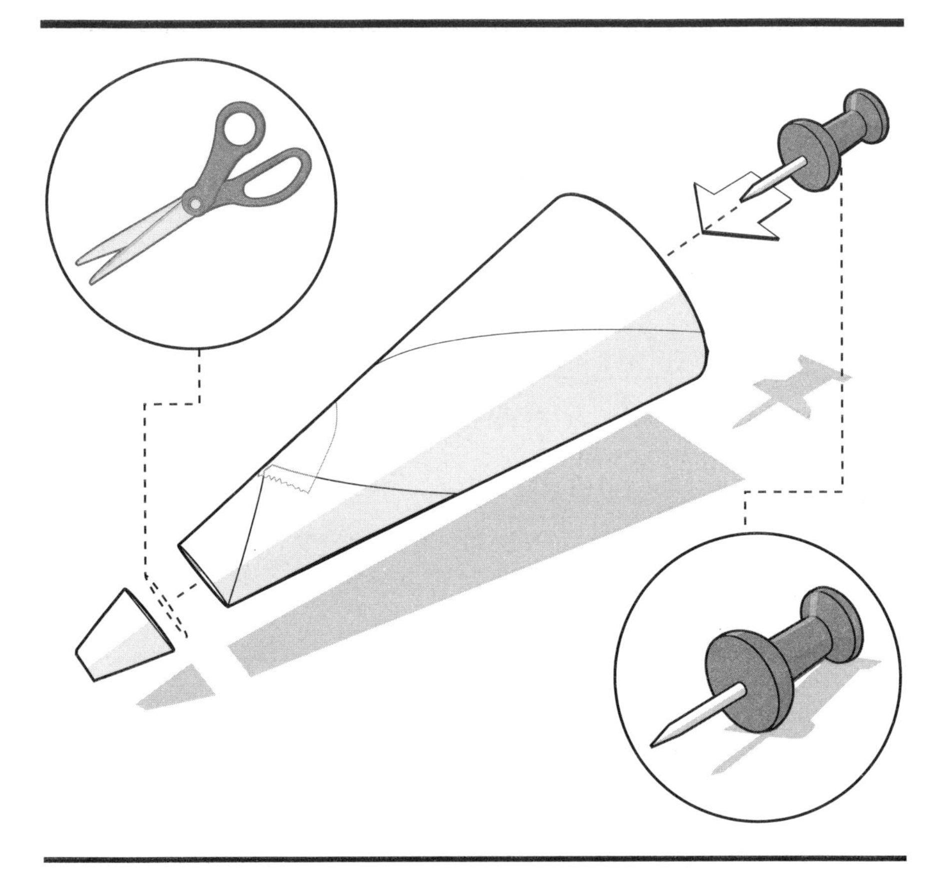

Remove the paper funnel from the tube. Trim off the nose of your paper funnel with scissors, just enough so that a pushpin will not fall out when placed inside.

Load a pushpin from the rear of the paper funnel with the metal point entering first. Tip the paper cone so that the pushpin slides all the way to the front of the cone. If the pin falls out of the front, your hole may be too large.

## Step 4

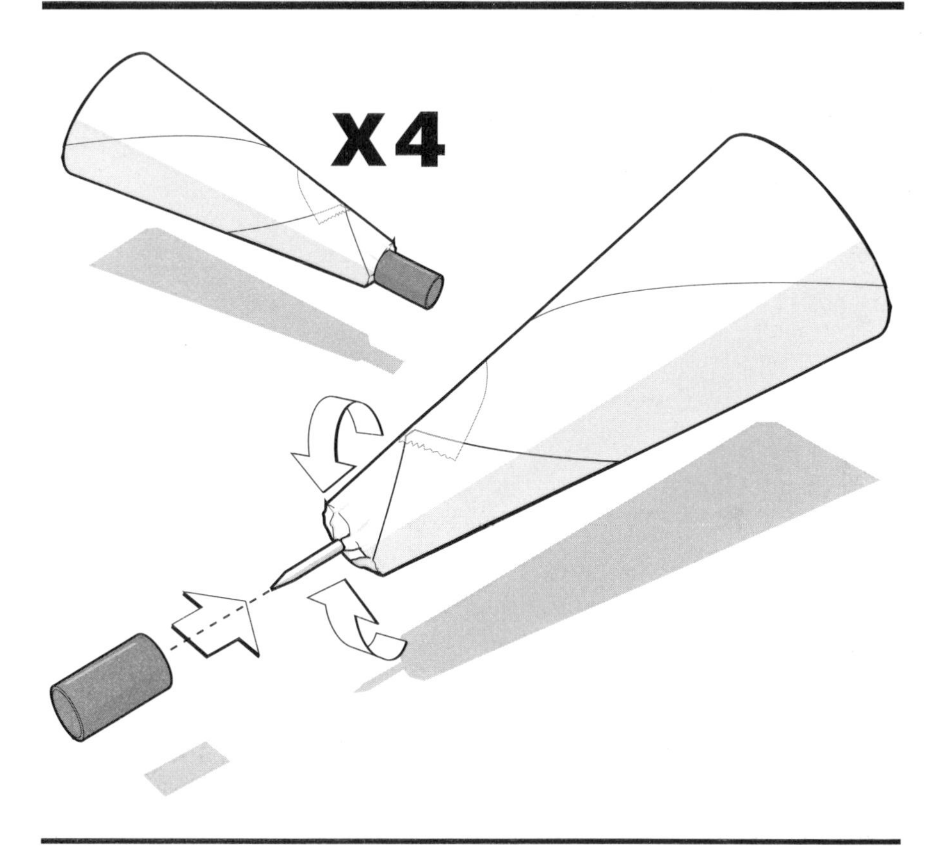


Cut the eraser off a pencil using a pocketknife.

With the tip of your pushpin protruding out of the paper cone, fold the paper inward around the metal point. Once all the paper is pushed down around the metal point, slide your pencil eraser onto the pin tip. Push the eraser as far back as possible to sandwich the pushed-down paper between the eraser and pushpin.

If you want multiple darts, repeat these steps until you've accumulated a stockpile of ammunition. This Paper Dart gun can hold four darts while firing one.

## Step 5

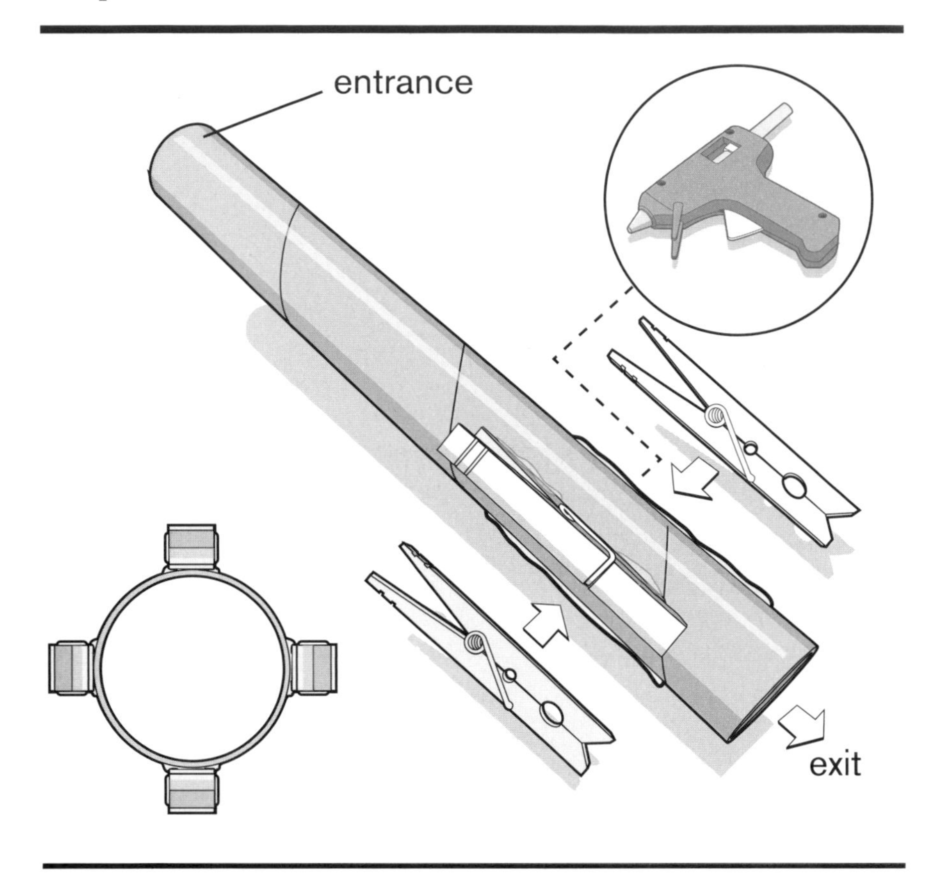


Your Paper Dart launcher is now fully functional without these additional steps; however, these improvements will impress your friends and frighten your enemies.

Using a hot glue gun, glue four clothespins to the firing end of your cardboard tube. They should be evenly spaced around the tube, an equal distance from the end. See the illustration for the pattern. Be careful to only glue the bottom wooden prong—the clip should still be functional.

# Step 6

During combat, the ammo clips will make it easier to safely transport your paper darts and will be a great reminder of how many shots you have remaining.

To fire, load one dart into the rear of the tube (eraser end first) and exhale into the tube as hard as you can with a quick burst of air. Your darts will be launched across the room with a surprising amount of accuracy.

Remember to always inspect your darts to make sure the pushpin is properly wedged into the eraser.

# 4

# CATAPULTS

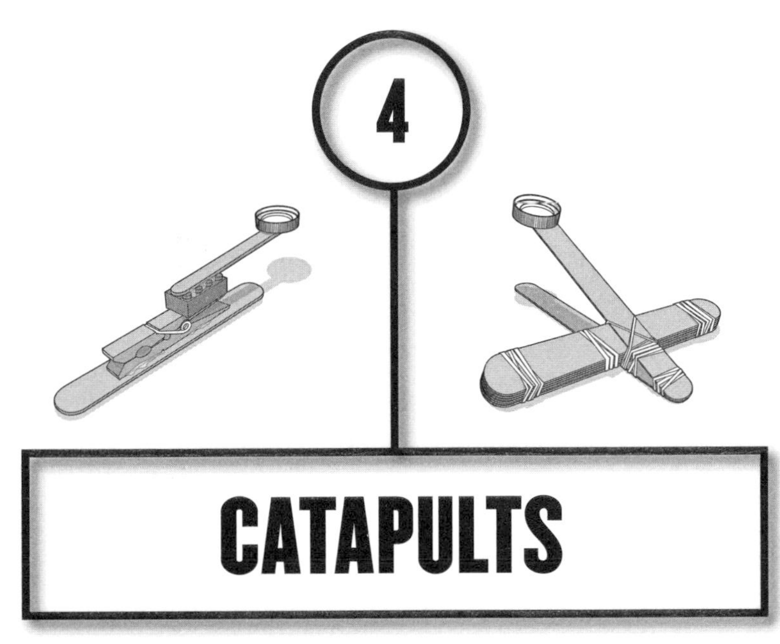

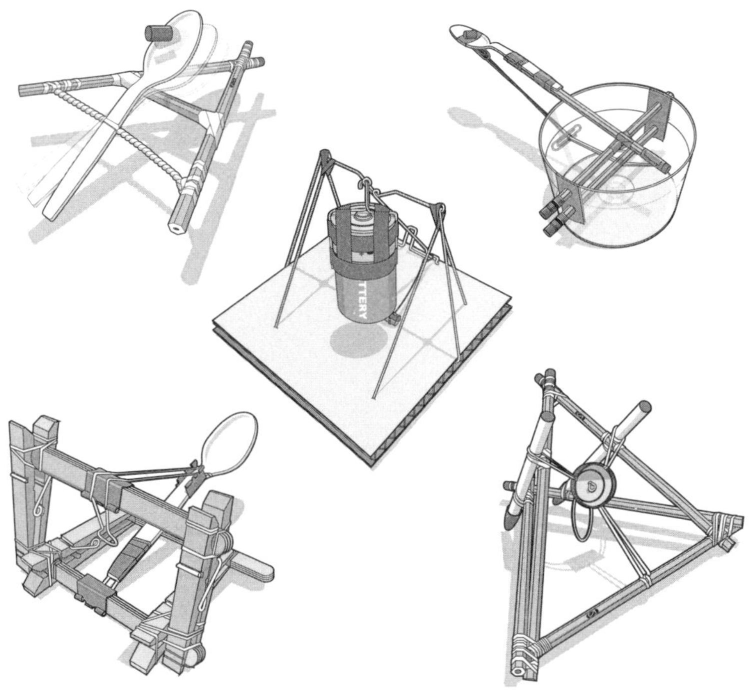

# CLOTHESPIN CATAPULT

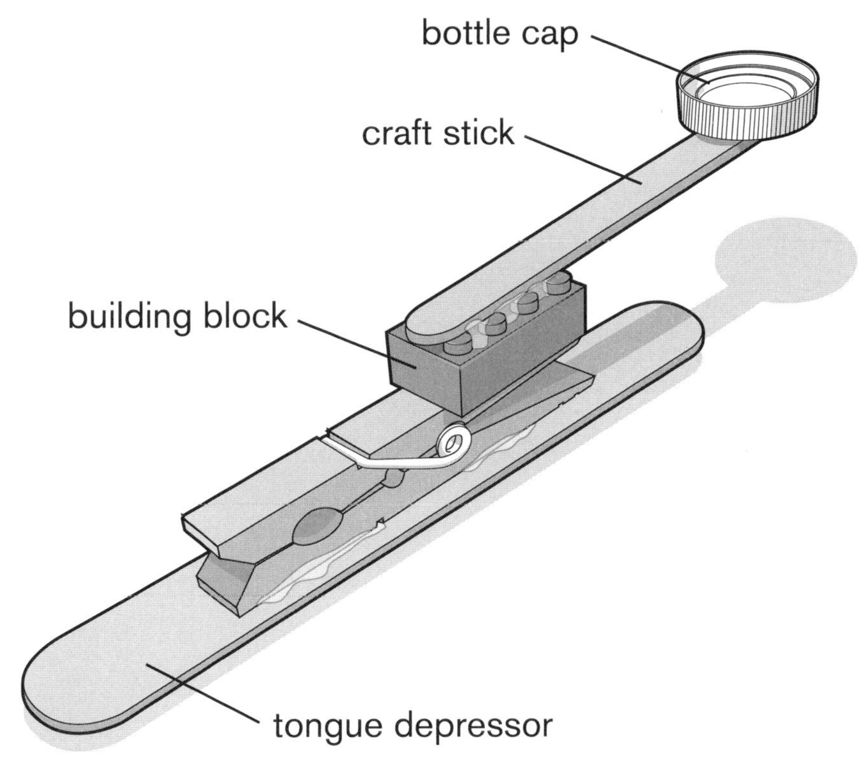


**Range: 10–20 feet**

This is a very simple catapult with endless possibilities for fun. It will only take a few seconds to glue and assemble, but finding a LEGO building block may take a bit longer. If you can't find one in the toy box, I suggest looking under the bed or behind the couch cushions.

### Supplies

1 toy building block
1 clothespin
1 tongue depressor
1 bottle cap
1 craft stick

### Tools

Safety glasses
Hot glue gun

### Ammo

1+ small marshmallows
1+ pencil erasers

## Step 1

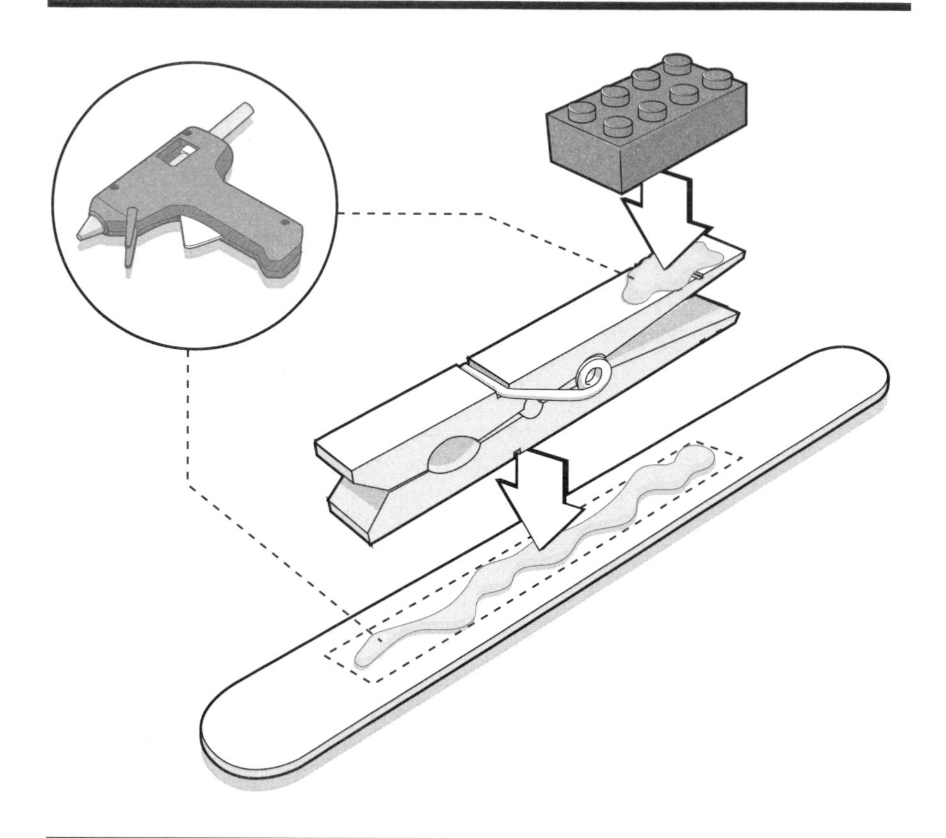

You'll need a 2-peg-by-4-peg LEGO building block (or something equivalent in mass and size) for this step. Using your hot glue gun, affix a clothespin to a wooden tongue depressor; then attach the toy brick onto the back end of the clothespin's top prong. The clothespin should remain fully functional.

Note that tongue depressors are easily found at your local craft or hobby shop, but they are almost impossible to find at pharmacies unless you preorder them online.

# Step 2

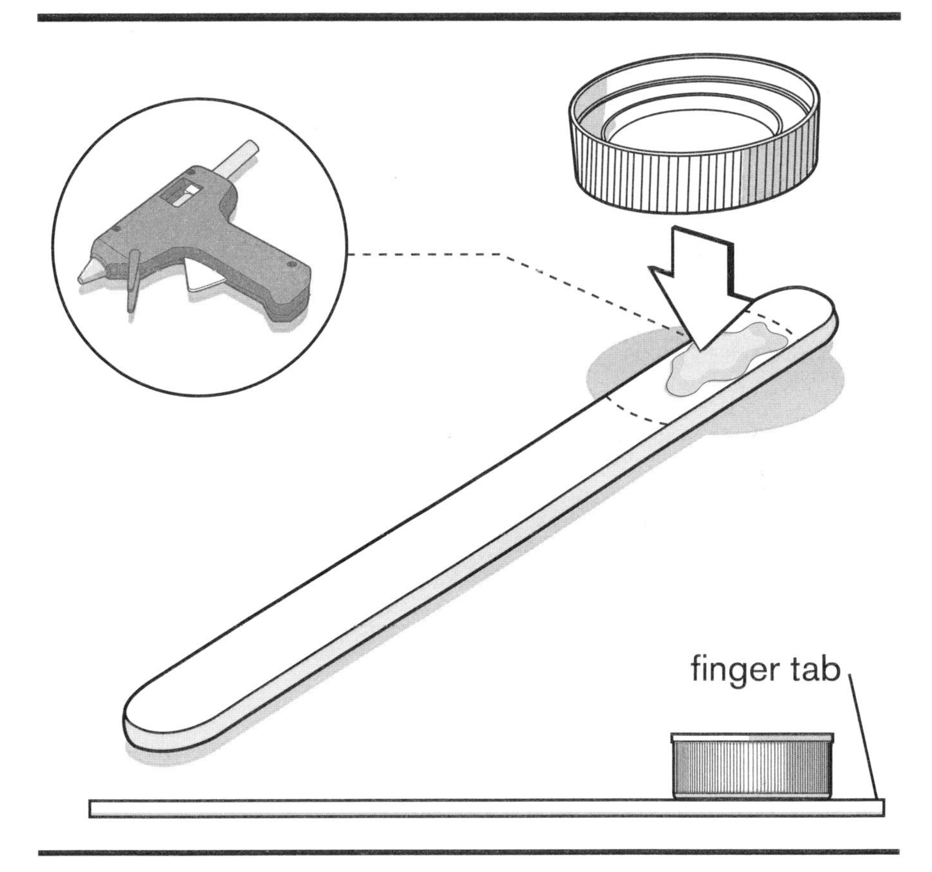


If you successfully operated the glue gun without burning yourself in the last step, here is another opportunity to do some damage. Glue a plastic bottle cap from a soft drink bottle to one end of a wooden craft stick, but leave a small tab at the end of the stick for your finger. This finger tab will come in handy when you begin your siege and have to quickly launch a hailstorm of ammunition.

## Step 3

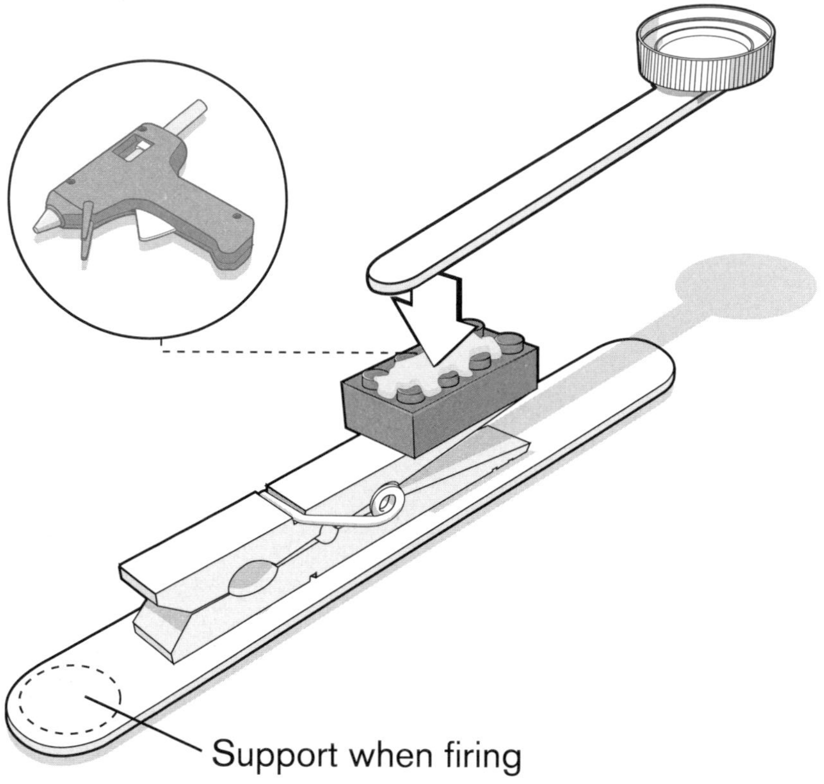

Support when firing

With your poor, burnt little fingers, combine the two individual assemblies to complete your Clothespin Catapult. Lay a small amount of hot glue onto the top of the building block and then carefully push the craft stick onto the brick.

Wait a few minutes for your glue to cool before test firing. Always wear safety glasses while operating a catapult. Small marshmallows or pencil erasers are the ideal ammunition for these contraptions. Substitution of ammunition could cause harm. Use at your own risk.

# DEPRESSOR CATAPULT

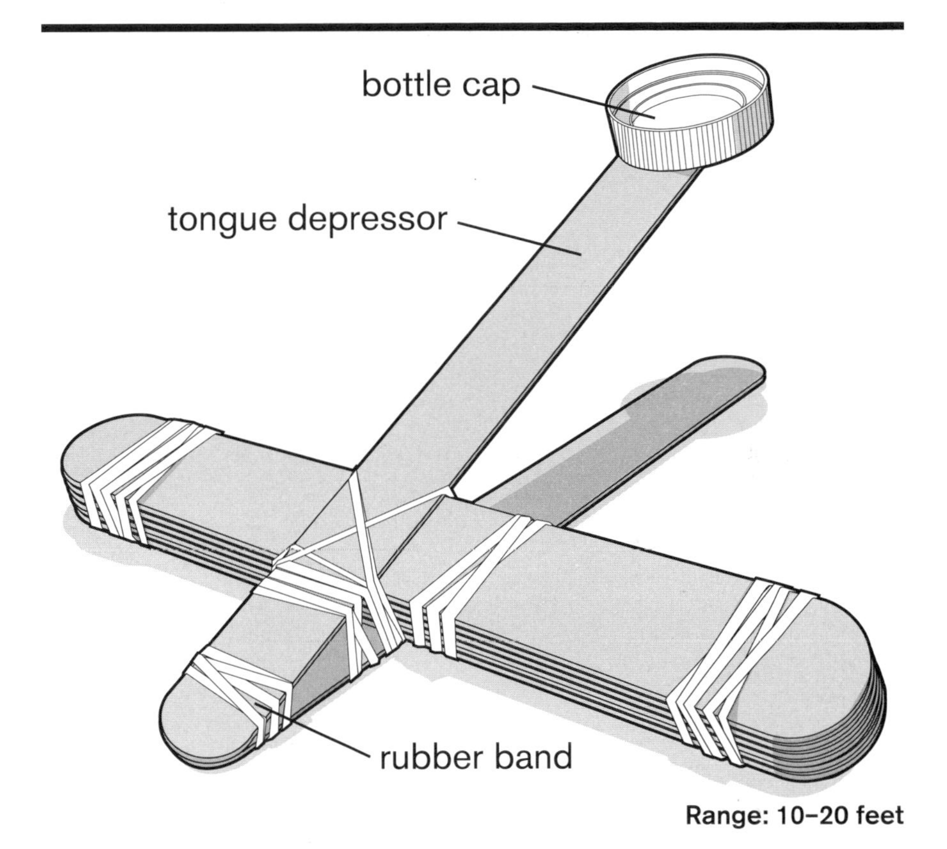


Range: 10–20 feet

This cut-rate catapult is perfect for mass production and outdoor use.

## Supplies

9 tongue depressors
7 rubber bands
1 bottle cap

## Ammo

1+ pencil erasers

## Tools

Safety glasses
Hot glue gun

## Step 1

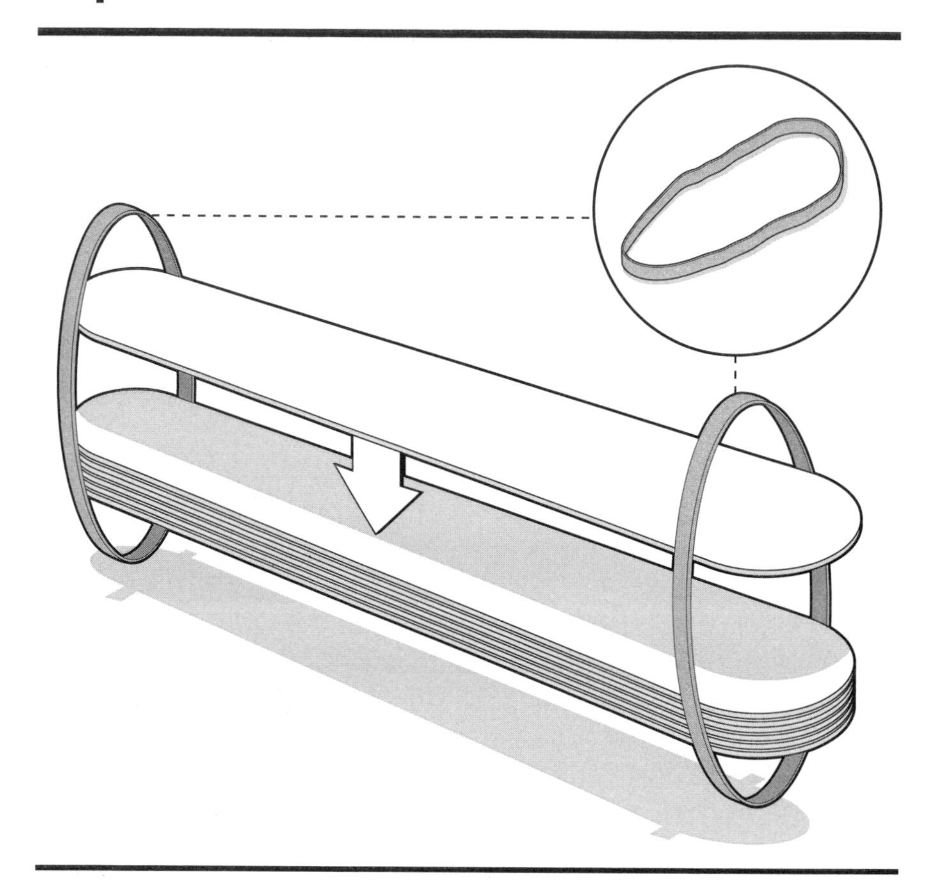

Seven wooden depressors total—that's how many you'll need in this step. Stack them on top of one another as neatly as possible before rubber banding them together at the ends.

Gluing them in this step would also work, though it will take some time.

## Step 2

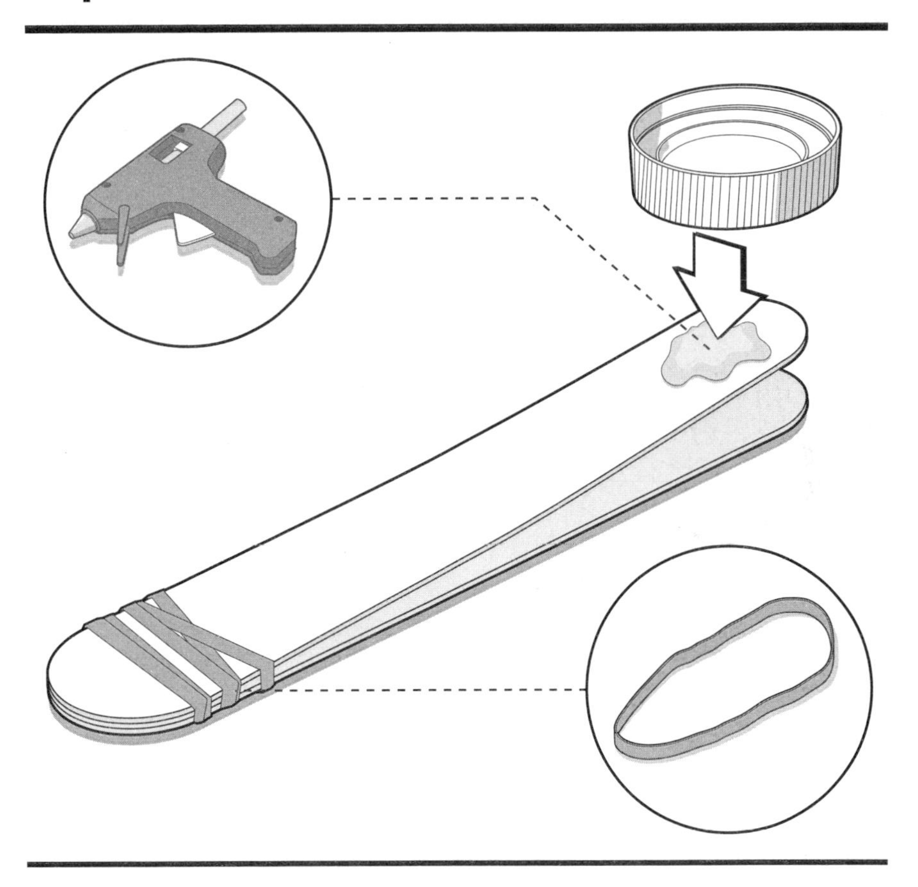

Take two additional tongue depressors and rubber band them together at one end.

Next, take a plastic cap from a soft drink bottle and hot glue it to the end of one of the tongue depressors, opposite the end held together by rubber bands. You have just finished your catapult arm.

Note: If you substitute craft sticks for tongue depressors they will eventually snap in half because of their thickness. They will work, but not for long.

## Step 3

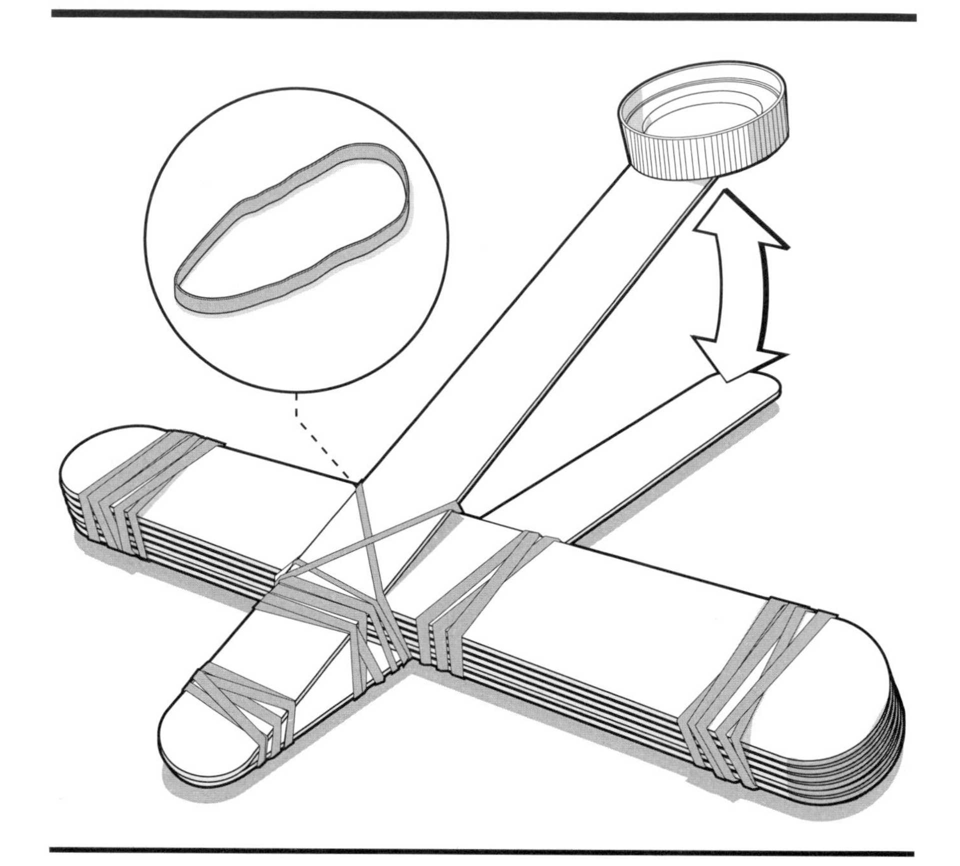

Before you can unleash catapulting terror on your targets, you must assemble the hinge arm. Slide the depressor bundle between the two tongue depressors from step 2 as shown. The tension of the arm assembly will force the bundle backward; use rubber bands to stop this and to hold it in place.

Now your catapult is complete. With one hand, hold the seven-depressor base on either side as you load the bottle cap. Pull down the top arm assembly to launch a soft projectile.

# #2 CATAPULT

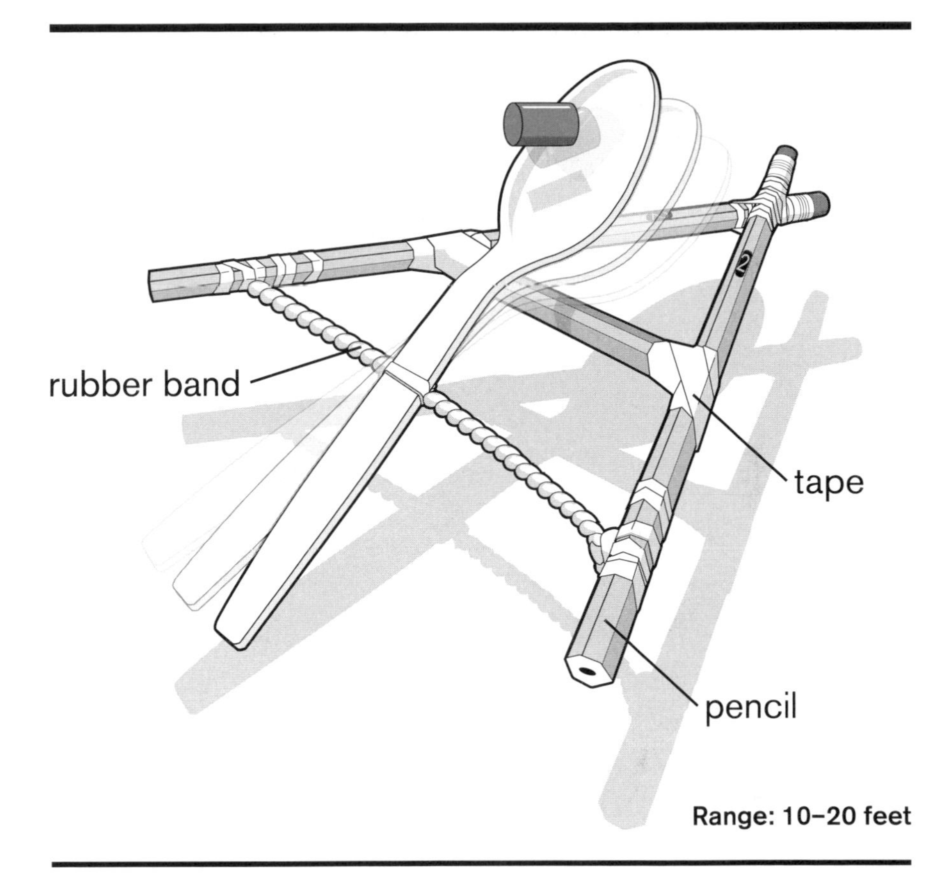


**Range: 10–20 feet**

The #2 Catapult is a great on-the-go siege machine. In order to fire it, you have to be holding it in your hand—no surface required. Once you've mastered the ideal trajectory angle, the catapult can be quite accurate.

## Supplies

3 wooden pencils
3 wide rubber bands
Masking or duct tape
1 plastic spoon

## Tools

Safety glasses
Pocketknife

## Ammo

1+ pencil erasers

## Step 1

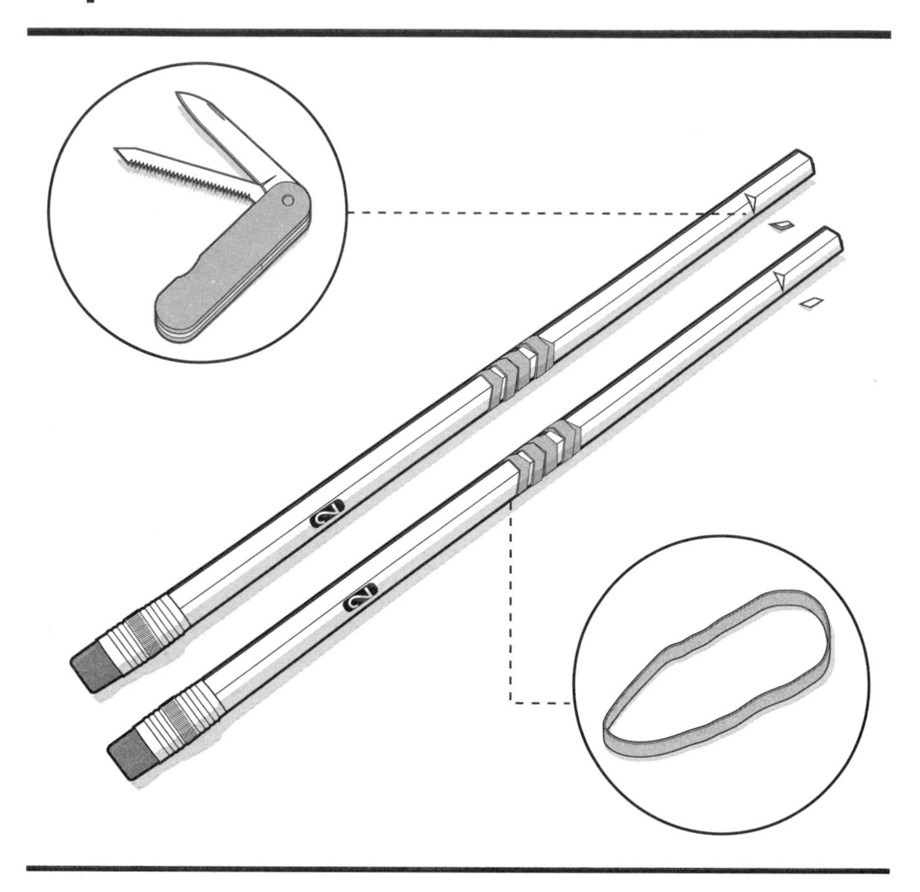

First, prepare two wooden pencils. Twist one rubber band around the center of each of the pencils until tight. You may have to first twist each band at the pencil end and then slide it down to the center.

Next, take a pocketknife and notch out two small groves about half an inch from the non-eraser ends, as shown.

## Step 2

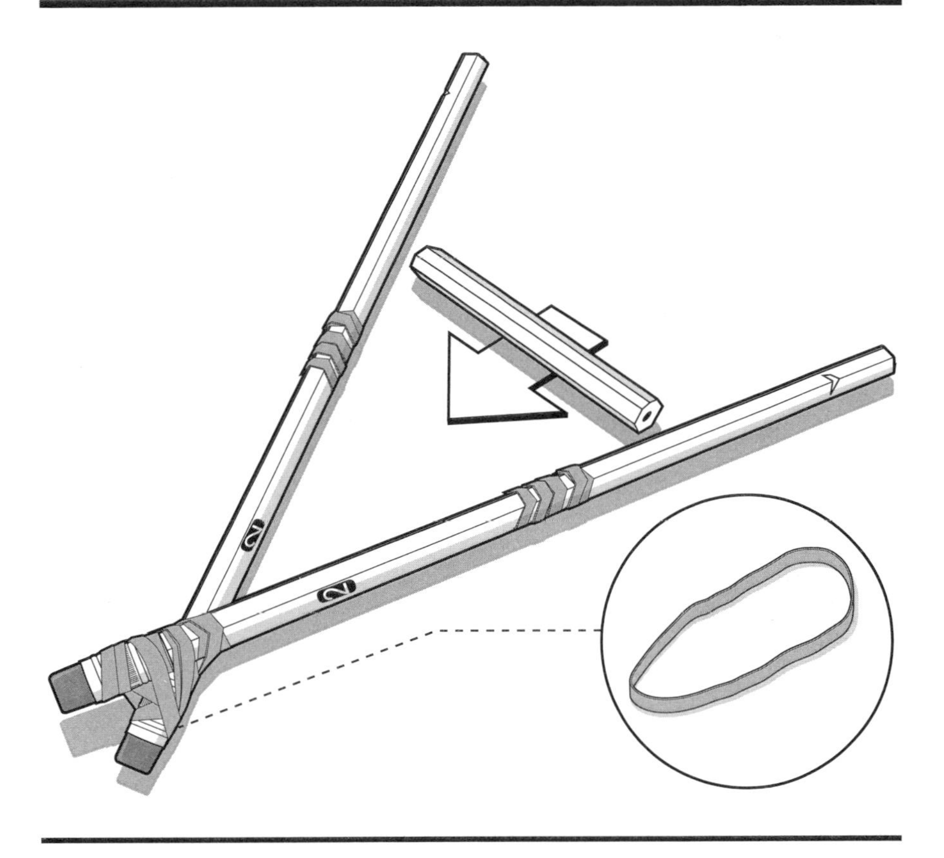

Attach the two modified pencils together with a rubber band at the eraser ends in a way that is flexible enough to create a V-shaped frame, as in the illustration above.

Take a third pencil and cut off the eraser. Then cut the rest of the pencil into two segments, one of which is 3 inches long. Clean up the cut edges with the pocketknife. The 3-inch segment will act as a spacer in the next step; set aside the remaining segment for later.

## Step 3

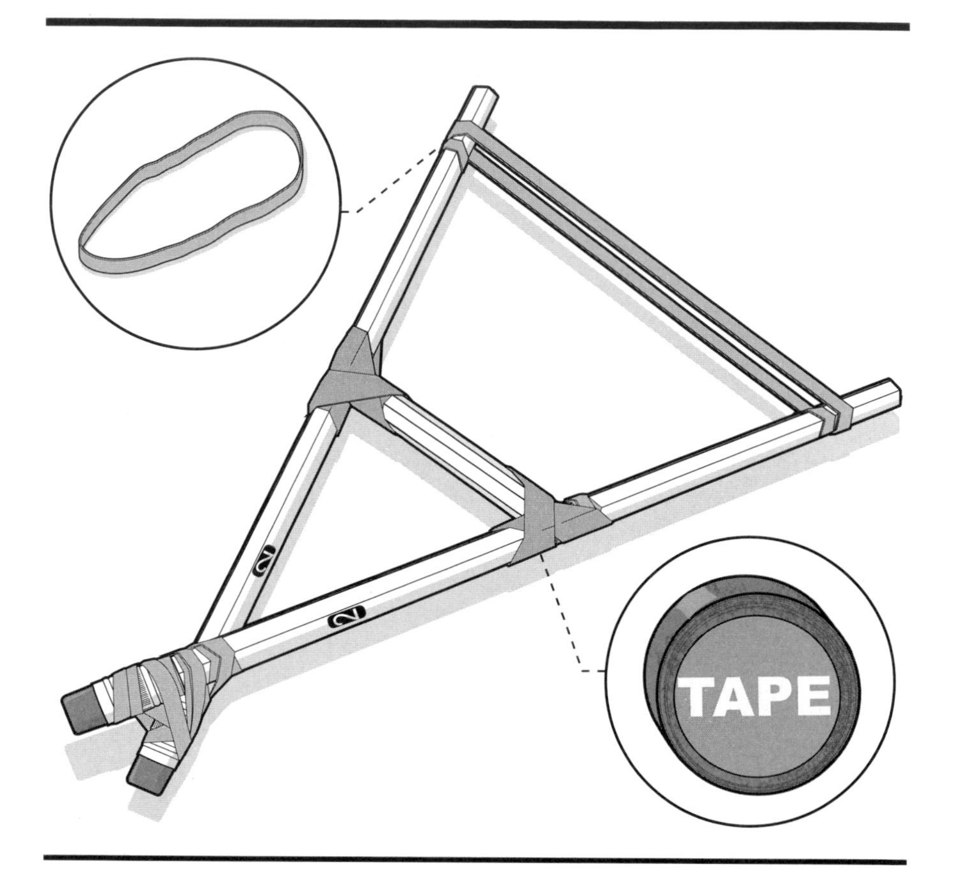


Tape this segment in place while it's resting on the rubber bands. The bands should add additional support and help the frame keep its structure.

Next, wrap one large, wide rubber band around the ends of the pencil near the two notches you cut out earlier. Just loop the rubber band a few times until it stays in place. Add additional rubber bands to hold it in place if needed.

## Step 4

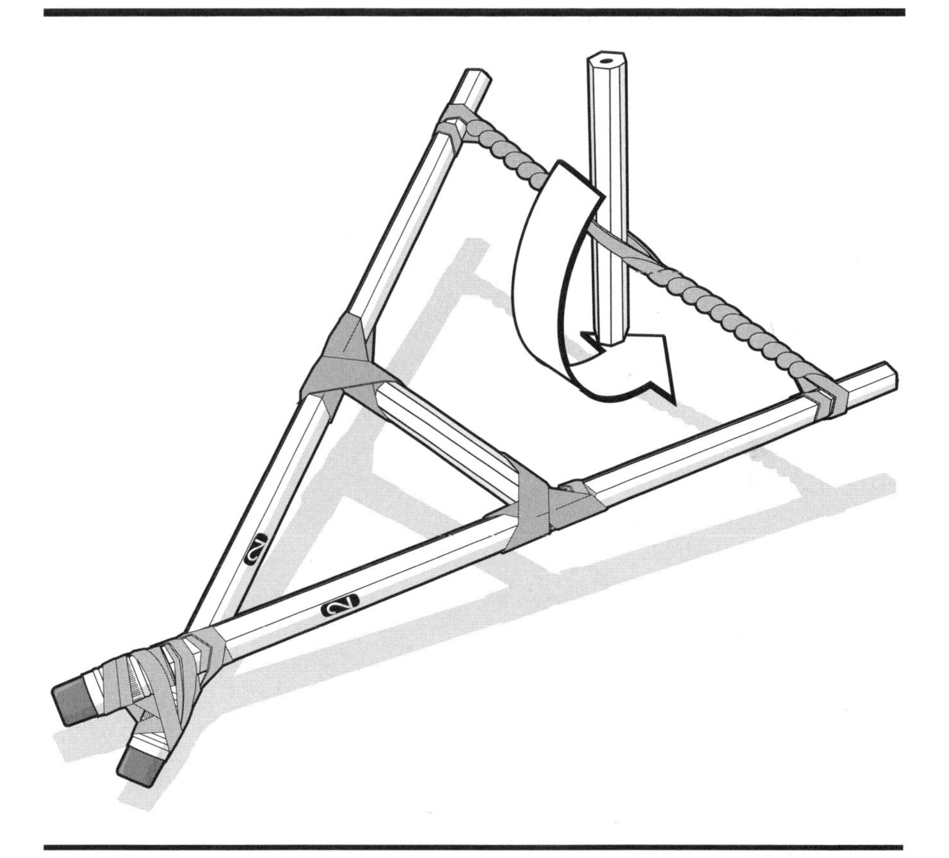

Now take the leftover section of the third pencil and slide it between the two sides of the thick rubber band you just installed. Once it's in place, use the pencil segment to spin and coil the rubber band. This stored energy will be the power source of the catapult.

Once you feel you've created enough elastic energy, ***do not let go!***

# Step 5

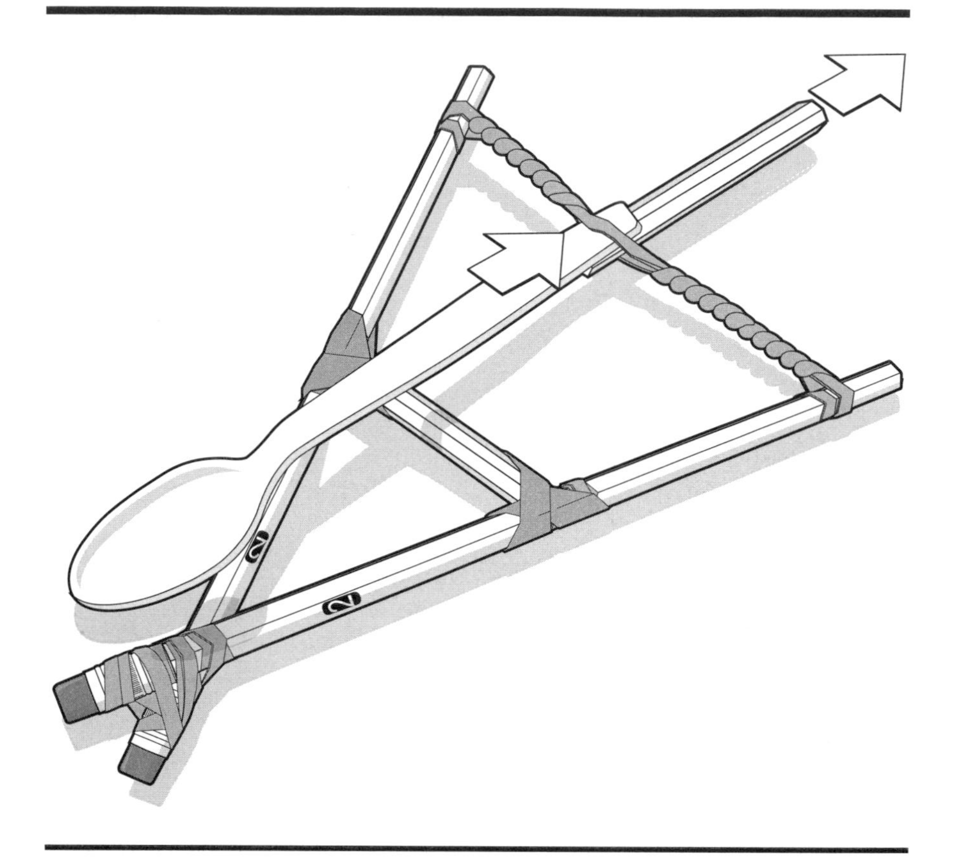

Now for the hard part. Slowly slide a plastic spoon handle through the rubber band opening, as shown. Once the spoon is in place, remove the pencil segment, and you're ready to go.

While holding the launcher, pull back the spoon, load an eraser, and let it rip. When operating, wear safety glasses and always prepare for the unexpected. Never fire ammo that can cause harm to another individual. If the rubber band is showing signs of wear, replace it before using the catapult again.

# CD-SPINDLE CATAPULT

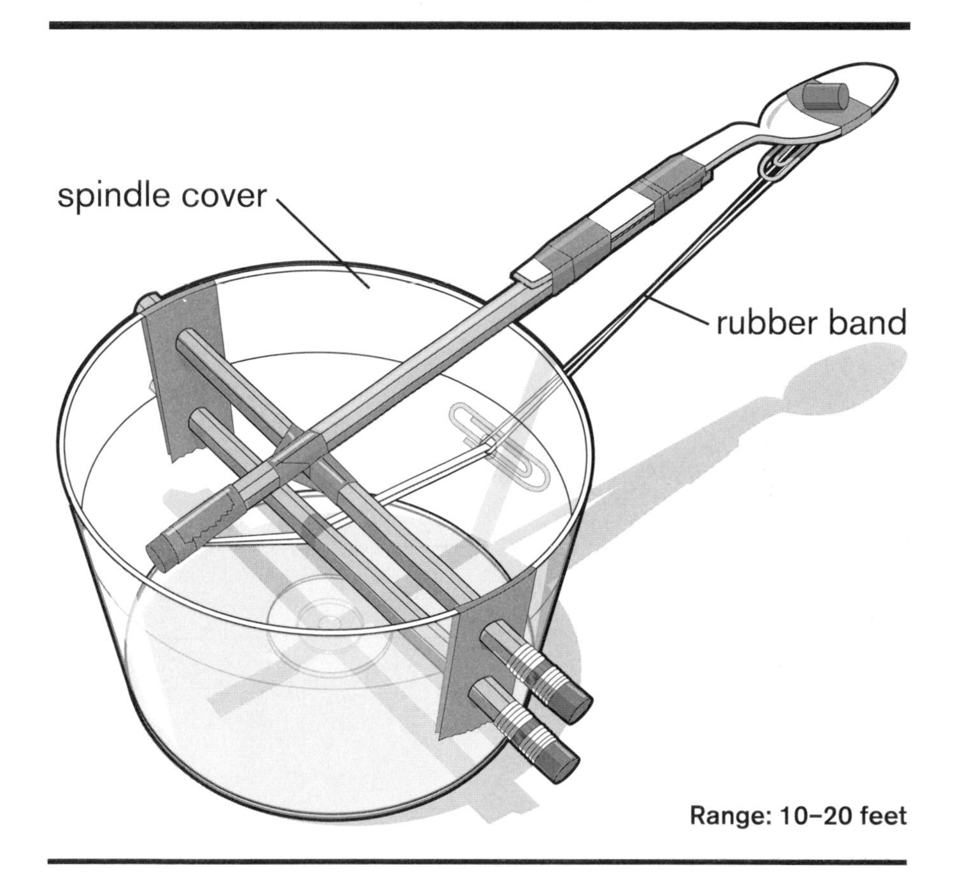


The CD-Spindle Catapult not only fires ammunition, but it also can store your shots as well. This project calls for a CD-spindle cap to use as the base; however, a shoebox or other smaller-sized containers will also work if a CD spindle is not available.

## Supplies

1 CD-spindle cover
Masking tape
2 paper clips
1 plastic spoon
3 wooden pencils
2 rubber bands

## Tools

Safety glasses
Hobby knife

## Ammo

1+ mini marshmallows

## Step 1

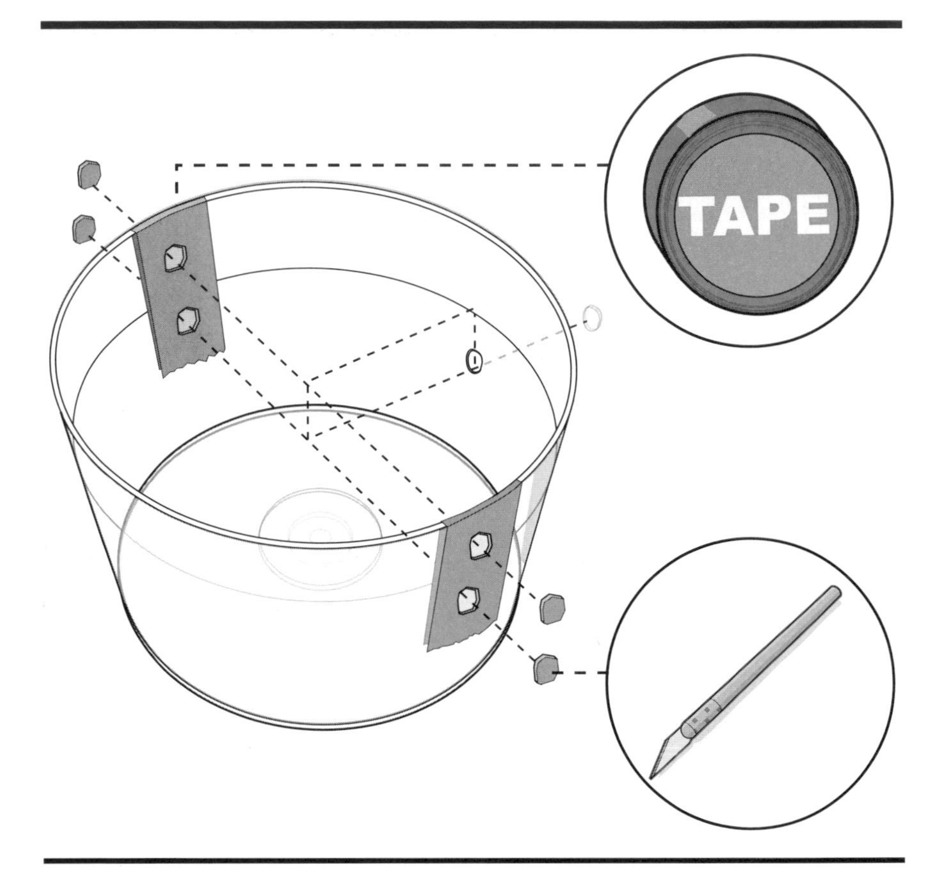


Take a CD-spindle cover and lay two pieces of masking tape directly across from one another on the inside and outside of the plastic surface. (This will help control the brittle plastic when you are cutting your holes.) Now take your hobby knife and cut a series of five pencil-sized holes into the plastic cover. Two sets should be exactly across from one another on the tape you stuck on earlier. The fifth hole should be cut midway between them at the same height as the bottom hole. Refer to the illustration above for the proper placement. Holes should not be cut close to the bottom.

## Step 2

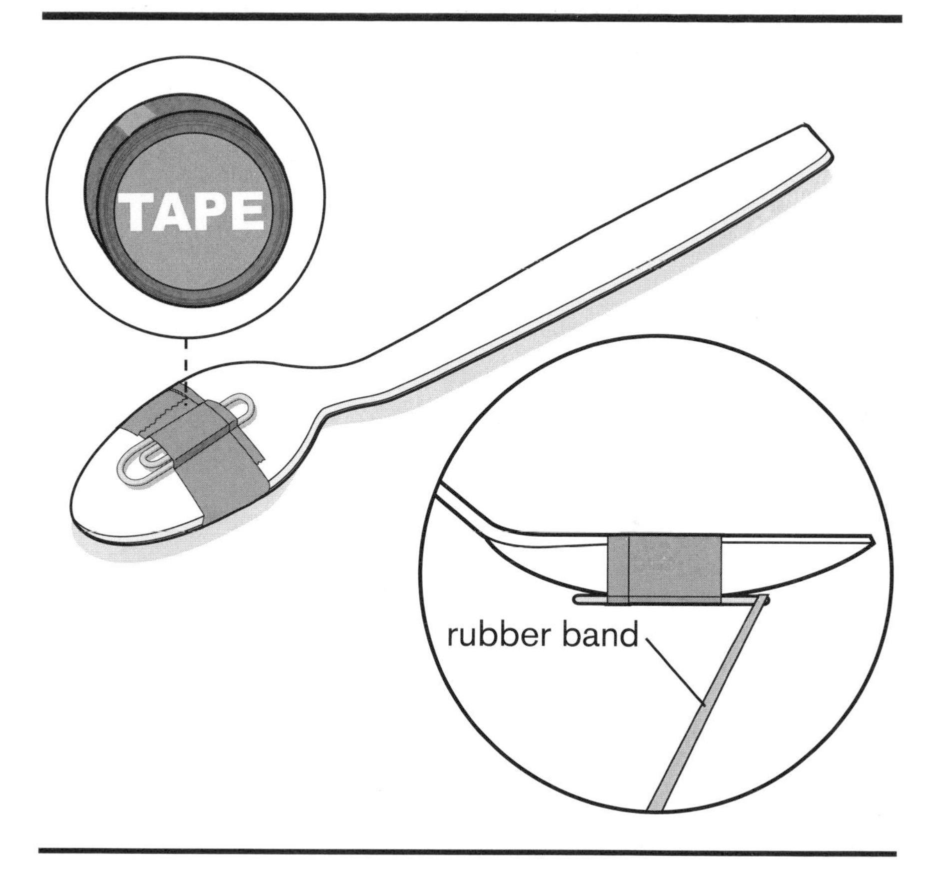


Next, construct the arm that will help you hurl objects across the room. Tape a standard paper clip to the belly of your plastic spoon. Later on, this paper clip will become your rubber band trigger, but you will not be putting a rubber band on it at this point. The illustration here is just for future reference.

## Step 3

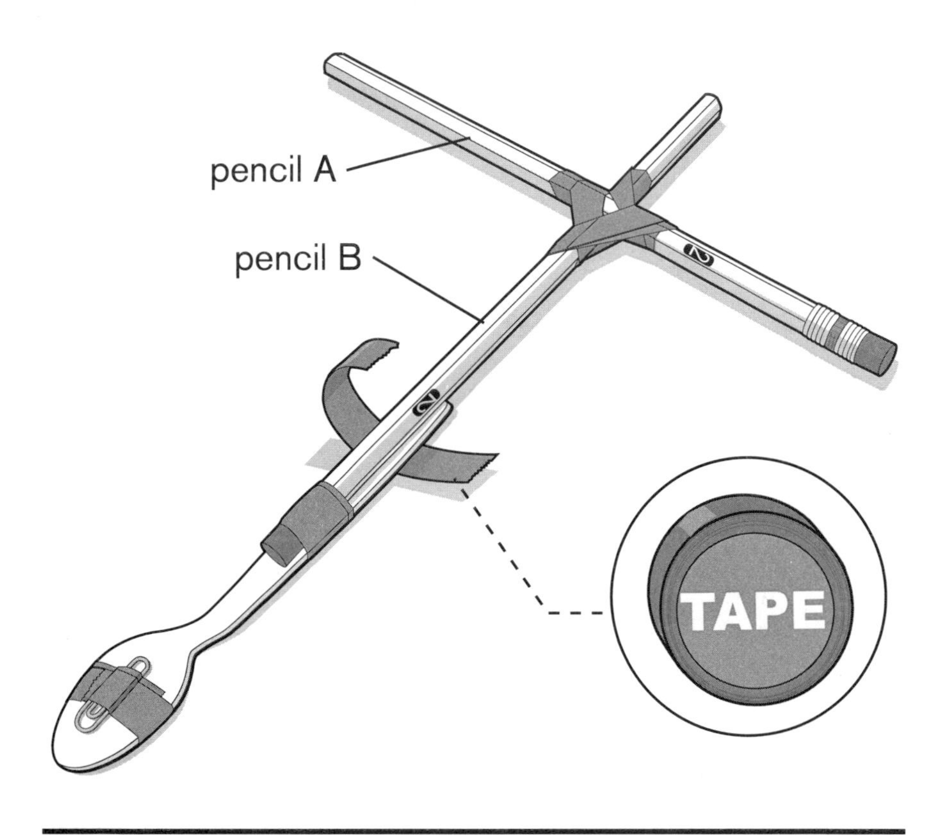


You will now complete the catapult arm with two wooden pencils and the spoon assembly from the previous step. Place pencil B in the center of pencil A (roughly three-quarters of the way down pencil B), as shown. Once in place, tape the two pencils together. You can substitute a rubber band for the tape, but be sure to refer to the illustration before construction.

Now fasten the plastic spoon onto the long end of pencil B. Tape in multiple spots to keep the spoon from becoming loose, which could reduce the power and accuracy of your catapult.

## Step 4

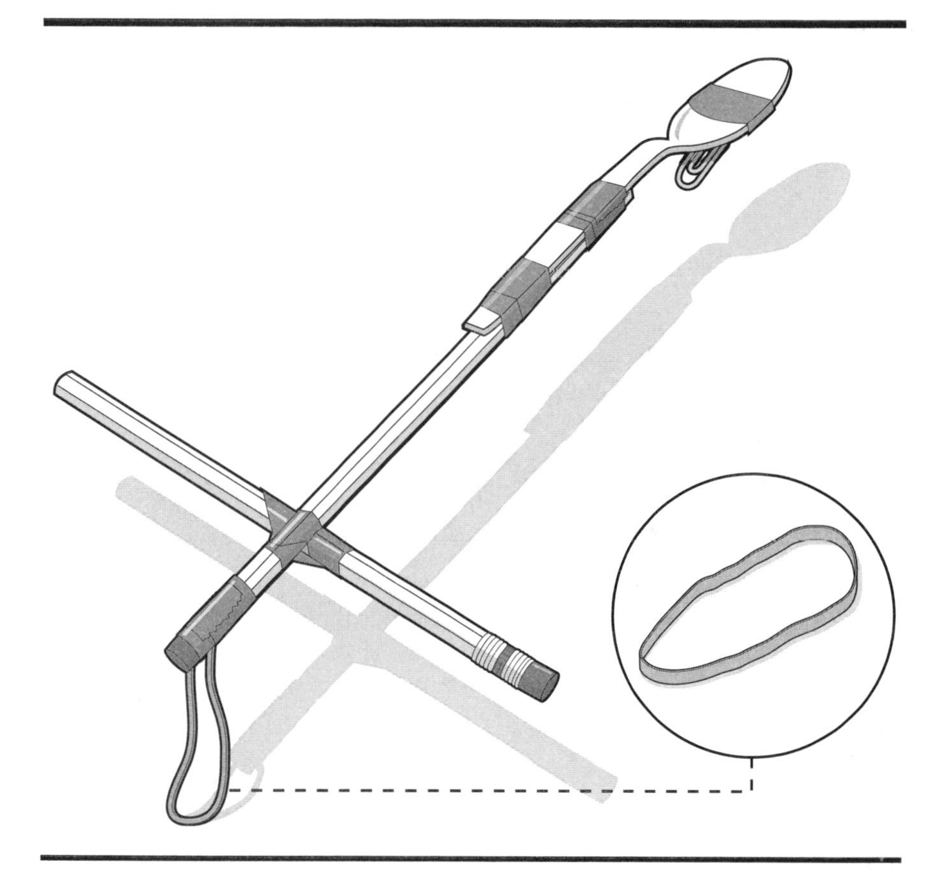

Tape a rubber band to the bottom of pencil B at the end opposite the spoon. This rubber band will provide the power to your catapult. The rubber band should be unused and without cracks.

This type of catapult is referred to as a torsion engine. A real torsion engine catapult would use treated twine or twisted rope to provide its wall-smashing power.

## Step 5

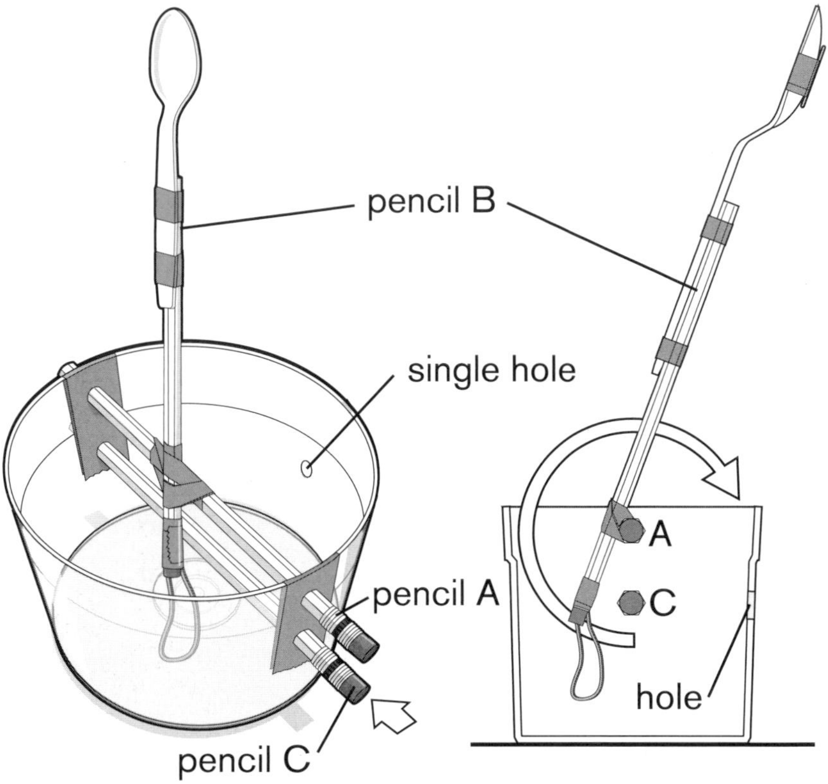

Now, from the inside of the spindle cover, delicately slide pencil A into the top two holes that you cut out earlier. Pencil B should not be touching the bottom of the container when vertical.

Next, slide in pencil C. This pencil will prevent the arm from swinging toward the ground, which would propel your ammunition into the ground as well.

Pencil B should be the farthest away from the single hole when vertical. See the illustration.

## Step 6

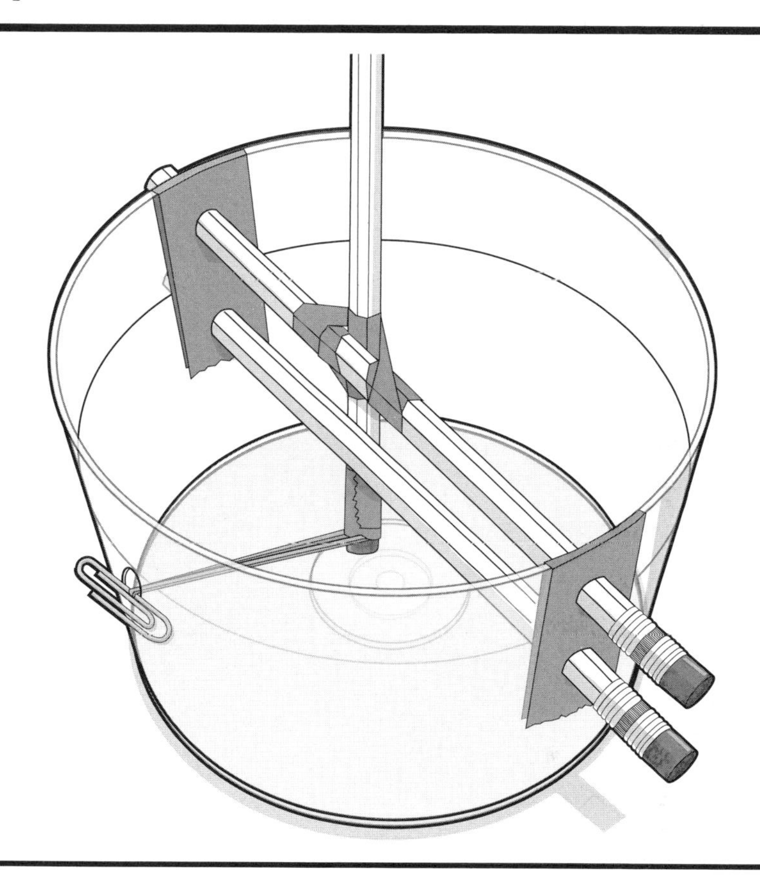

Now looking at the catapult from the opposite side. Pull the rubber band attached to pencil B through the single hole and secure it to a paper clip on the other side of the plastic. This will prevent the rubber band from breaking free and disabling your siege weapon.

Your catapult should work at this stage; however, you will find it difficult to hold, load, and fire all at the same time. The next step illustrates how to use the rubber band trigger you installed earlier on the plastic spoon.

# Step 7

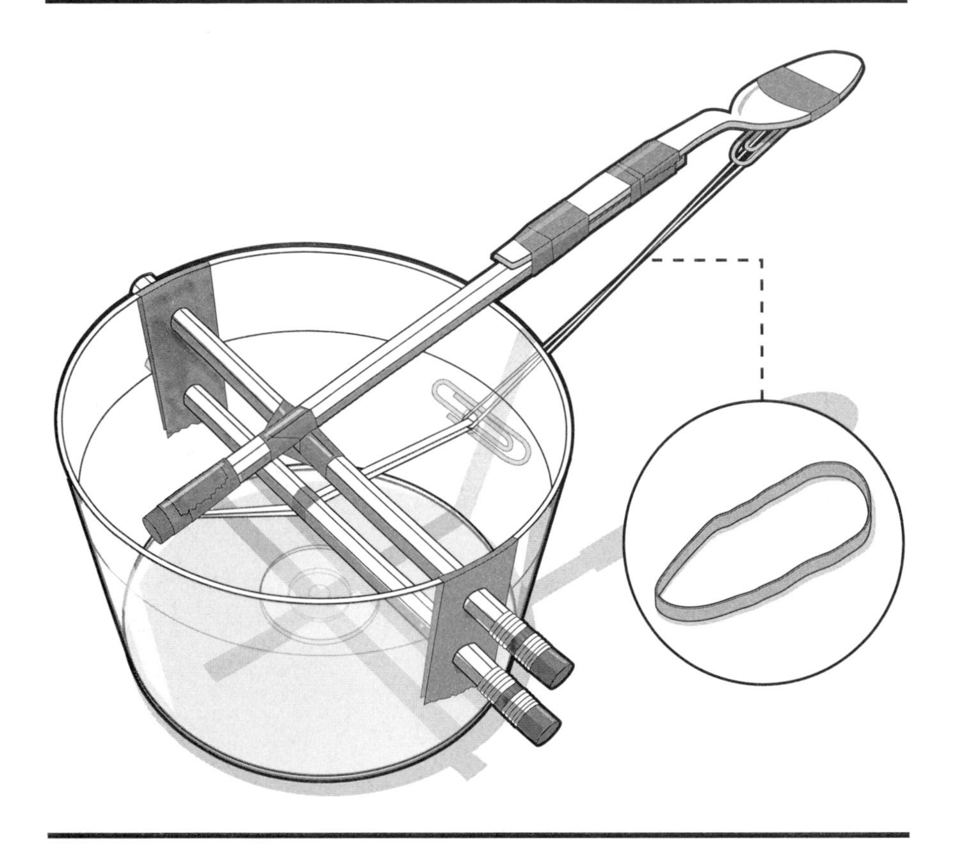

Loop a rubber band around both paper clips you installed earlier. This will keep the catapult in the locked position, perfect for loading and aiming. Once loaded, use your finger to flick the rubber band connection off the spoon. Just remember to use your other hand to support the base when firing.

The CD spindle's cover works great during mini marshmallow battles. Use your opponents' cover for a scoring basket when testing your skills.

# SIEGE CATAPULT

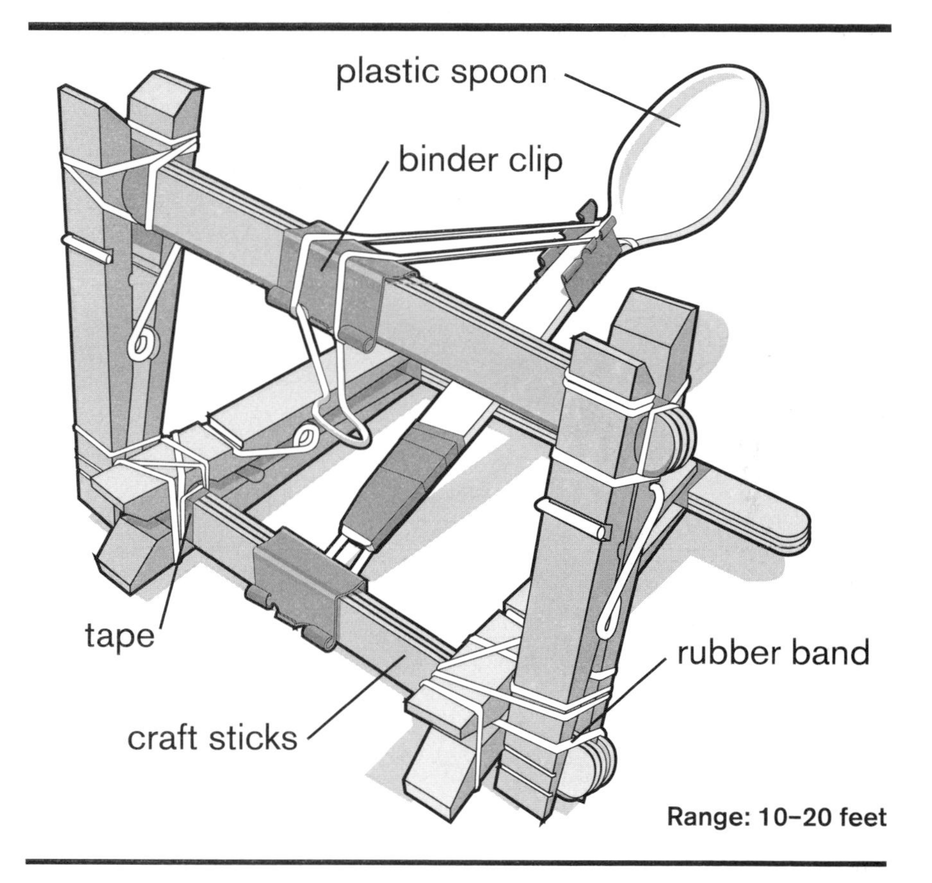


**Range: 10–20 feet**

Not only is this a true torsion engine catapult, it even looks like its real-life big brother. This catapult can be built for around 50 cents, which makes it great for producing in mass quantities.

## Supplies

9 craft sticks
Masking or duct tape
4 clothespins
7+ rubber bands
3 small binder clips (19 mm)
1 plastic spoon

## Tools

Safety glasses

## Ammo

1+ mini marshmallows

## Step 1

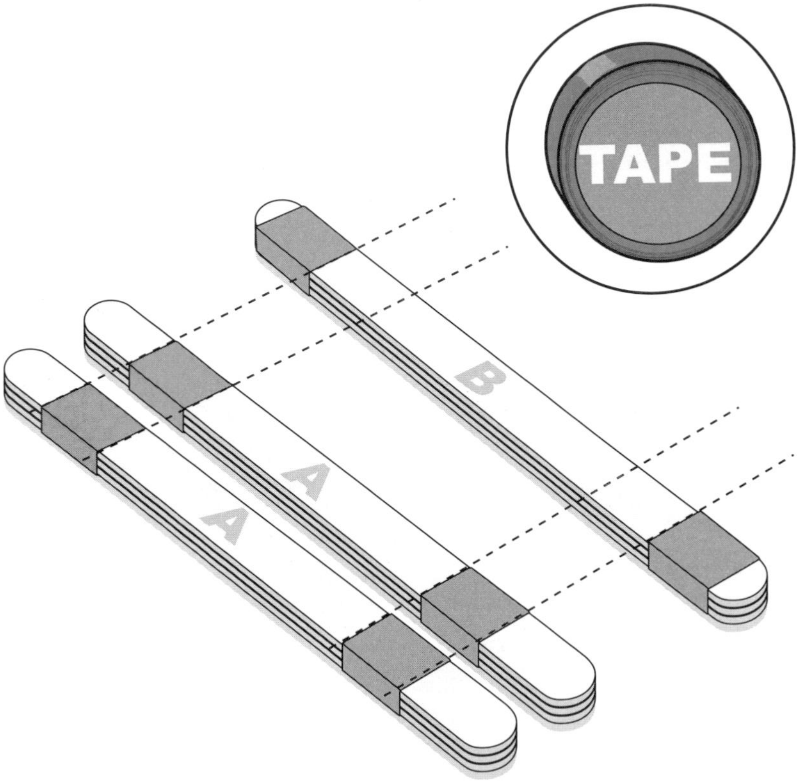

First, bundle up your craft sticks. Take nine sticks and divide them up into three piles—two *A* piles and one *B* pile. Each pile should be three sticks high.

Begin by taping both A piles about a half an inch from their ends. Pile B should be taped just before the round ends on the sticks, as illustrated.

## Step 2

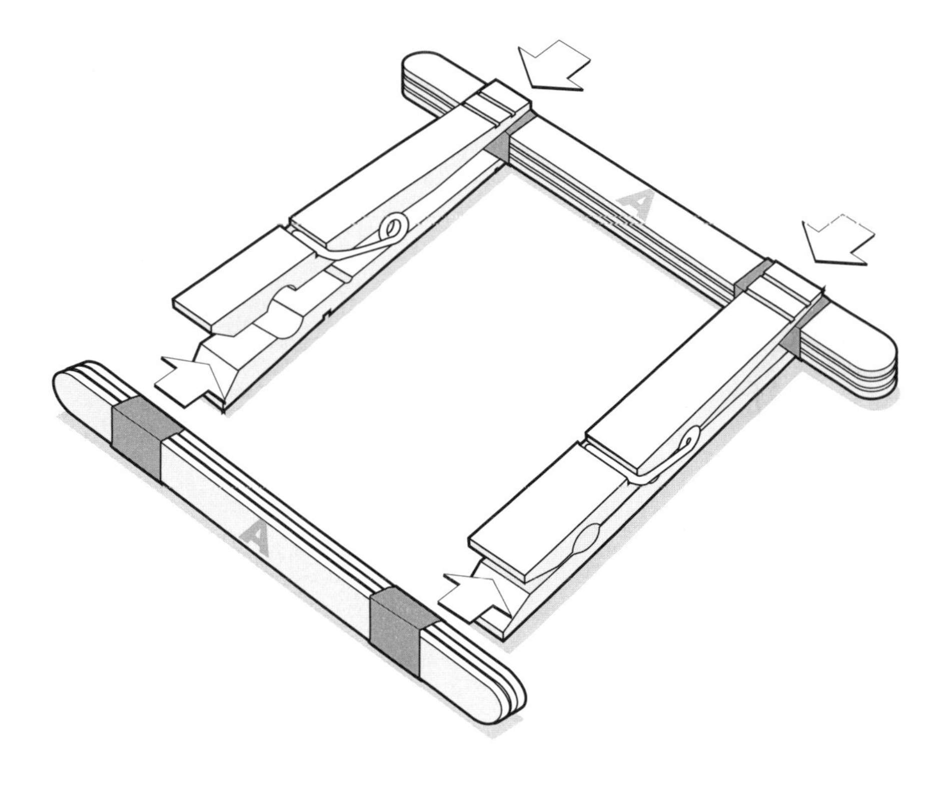


Next, take one of the A piles and clip it into two wooden clothespins. The clothespins should be positioned on the tape, and the craft sticks should be vertical. Then slide the second A pile horizontally between the clothespin prongs at the rear of the clothespins. Refer to the illustration.

The taped areas should help hold the final frame in place when the catapult is being operated.

# Step 3

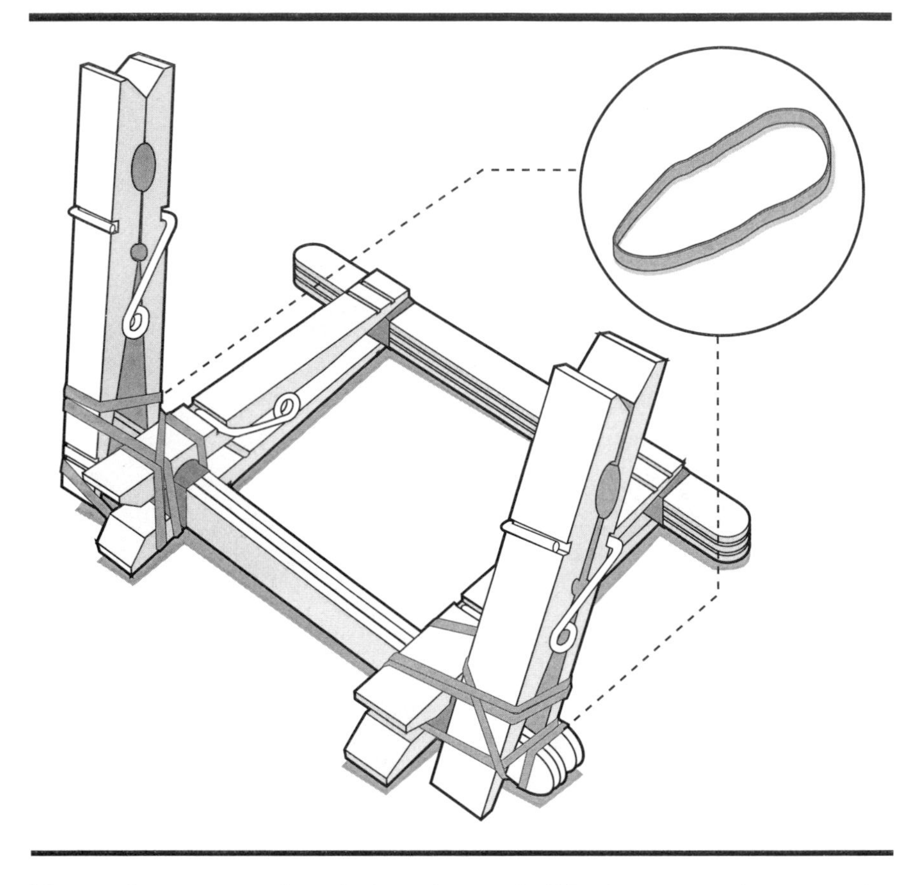

Next, take two wooden clothespins and slide the prongs between the craft sticks in the vertical bundle as illustrated. You'll have to muscle the craft sticks apart in order to fit each clothespin prong between them.

Now strap the two clothespins in place with a rubber band at each end. Once you feel the clothespins are secure, move on to the next step. If you want to add additional rubber band support to the rear, please do so.

## Step 4

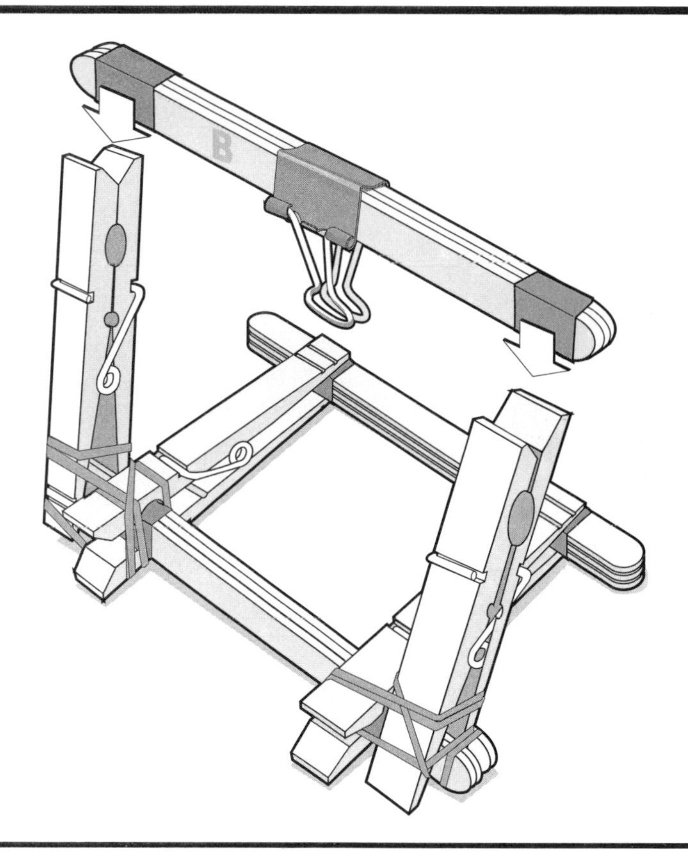


Take a small binder clip and clamp it to the center of the third bundle of sticks—pile B. This is the pile with the tape at the ends.

Now clip the bundle into the upright clothespins, making sure to line the clothespins up with the tape. The craft sticks in the bundle should be vertical and the binder clip's metal handles should face down, as shown.

## Step 5

Now it's time to construct the plastic hurling arm with two more small binder clips and a plastic spoon.

First, take a binder clip and tape one of its metal handles to the underside of the plastic spoon, as shown. (Do not remove the second metal handle on this clip.)

Next, attach another binder clip to the underside of the spoon's neck. Squeeze both metal handles from the sides to release. (Binder clips vary by manufacturer, so if you are unable to remove the handles, you may have to skip attaching the binder clip to the neck and instead tape the rubber band onto the spoon in step 7.)

## Step 6

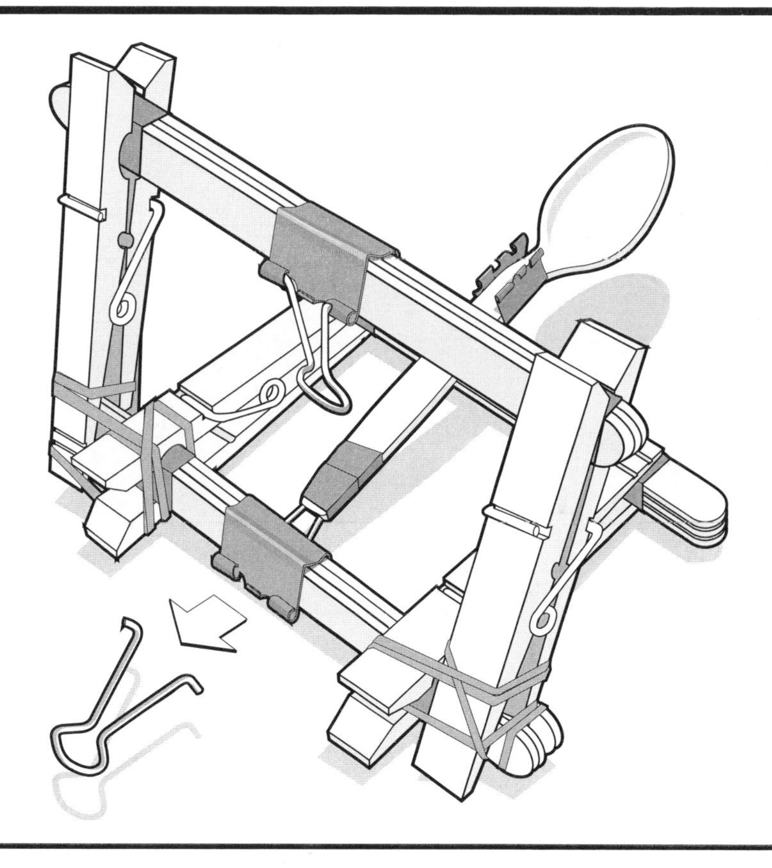

Now, use the binder clip attached to the spoon handle to clip the spoon onto the lower craft-stick brace. The clip should be centered on the brace. Once the spoon is in place, remove the front metal handle of the binder clip.

## Step 7

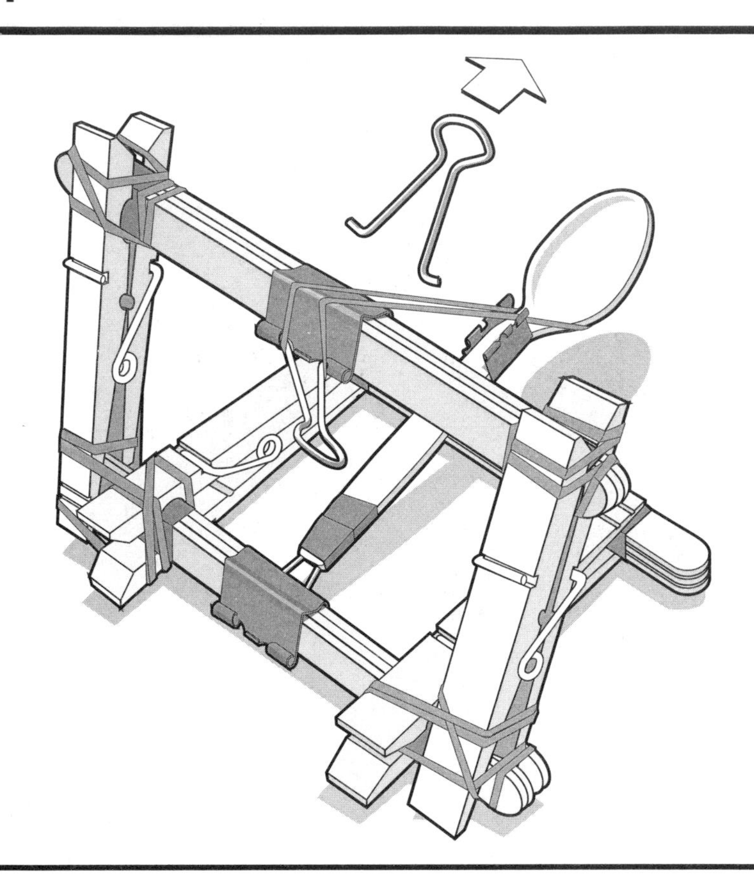

Now loop a small rubber band around the neck of the plastic spoon. Position the rubber band so it runs through the clip attached to the spoon's neck. This clip will hold the rubber band into position.

Loop the other end of the rubber band around the binder clip located on the top craft-stick bundle. Wedge it under the metal handle on the clip, and if necessary, clip it under the binder clip to hold it in place. Now remove the back metal handle on this clip (the handle on the side facing the plastic spoon), and you're ready to launch.

Remember the importance of safety when operating your Siege Catapult. Never aim it at another human or animal and only use safe ammunition. Mini marshmallows work nicely.

# VIKING CATAPULT

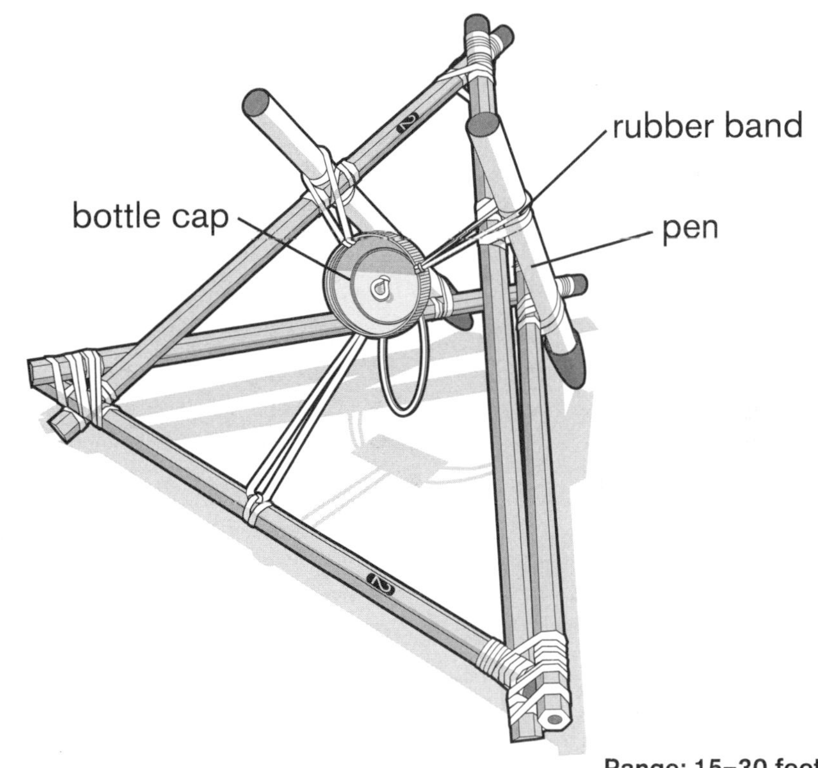


**Range: 15–30 feet**

The Viking Catapult is a very practical tabletop siege weapon with unlimited ammo possibilities. It's capable of launching coins, spitballs, erasers, marshmallows, and much more. With an intimidating silhouette and a straightforward firing mechanism, it's the perfect machine to inflict physiological and physical damage on any annoying target.

### Supplies

5 wooden pencils
14+ rubber bands
2 plastic pens
1 bottle cap

### Tools

Safety glasses
Hobby knife

### Ammo

1+ pencil erasers (can be taken from the frame's pencils)

## Step 1

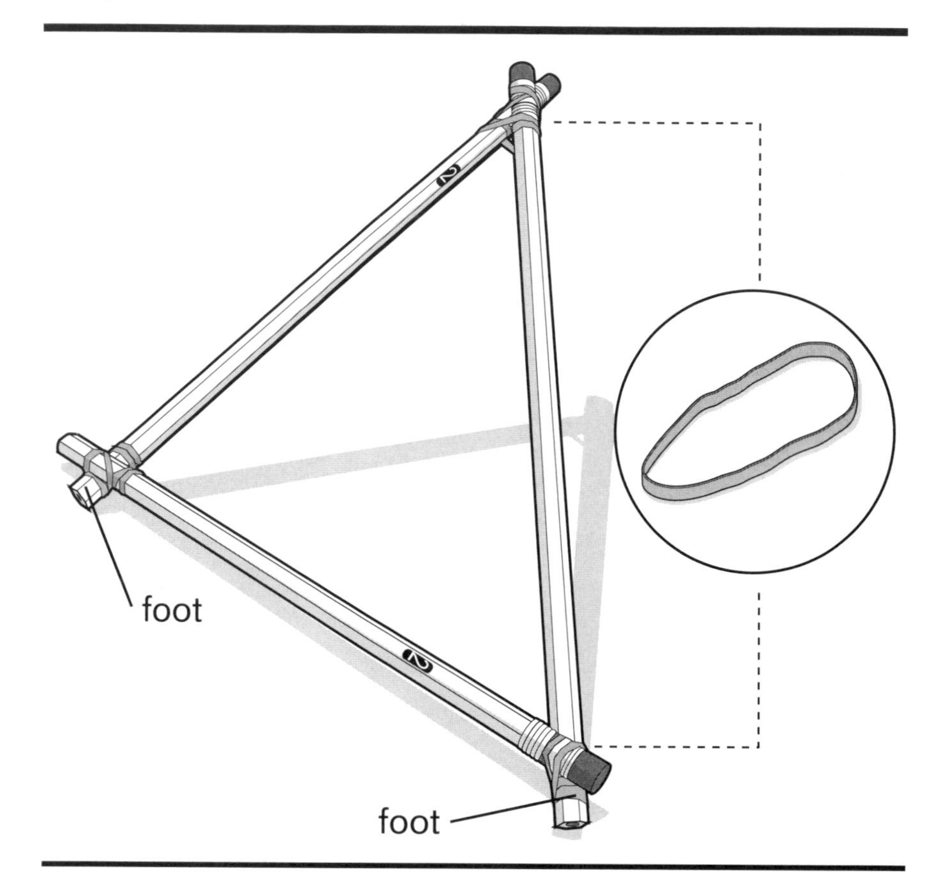


First, rummage through your desk drawers and select three wooden pencils. Use them to construct a triangular frame, securing the connections with rubber bands. You will have some overhang at the ends; these will become your feet.

## Step 2

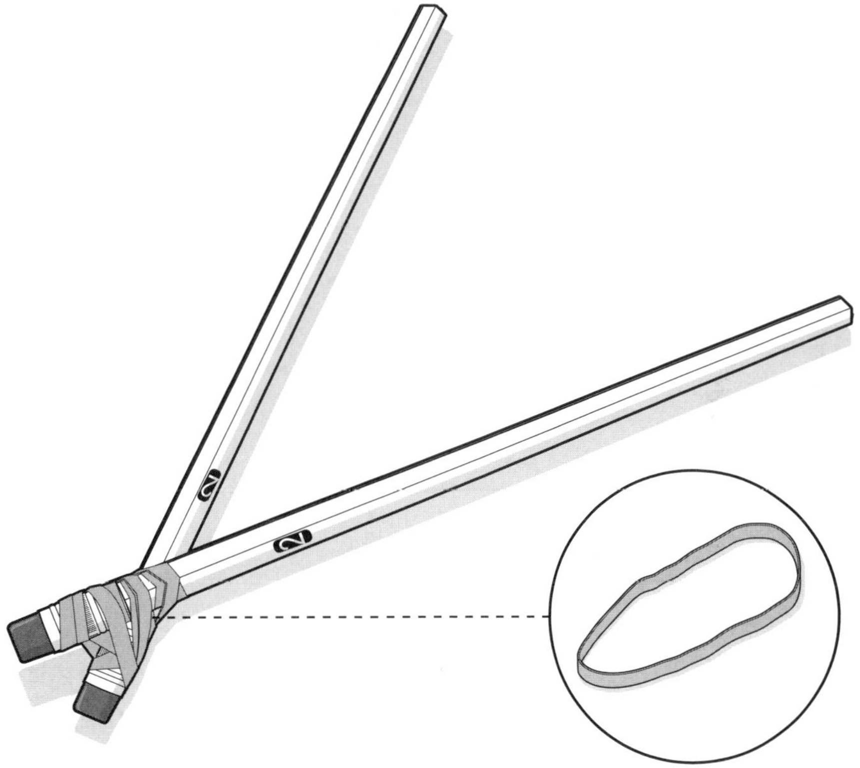

Next, take two more pencils and lash them together at one end using a rubber band or two, similar to the construction of the #2 Catapult you built earlier (page 129).

The connection should be tight enough to hold the pencils together but loose enough to bend them into a V shape for the next step. This assembly will make up half of the Viking Catapult.

## Step 3

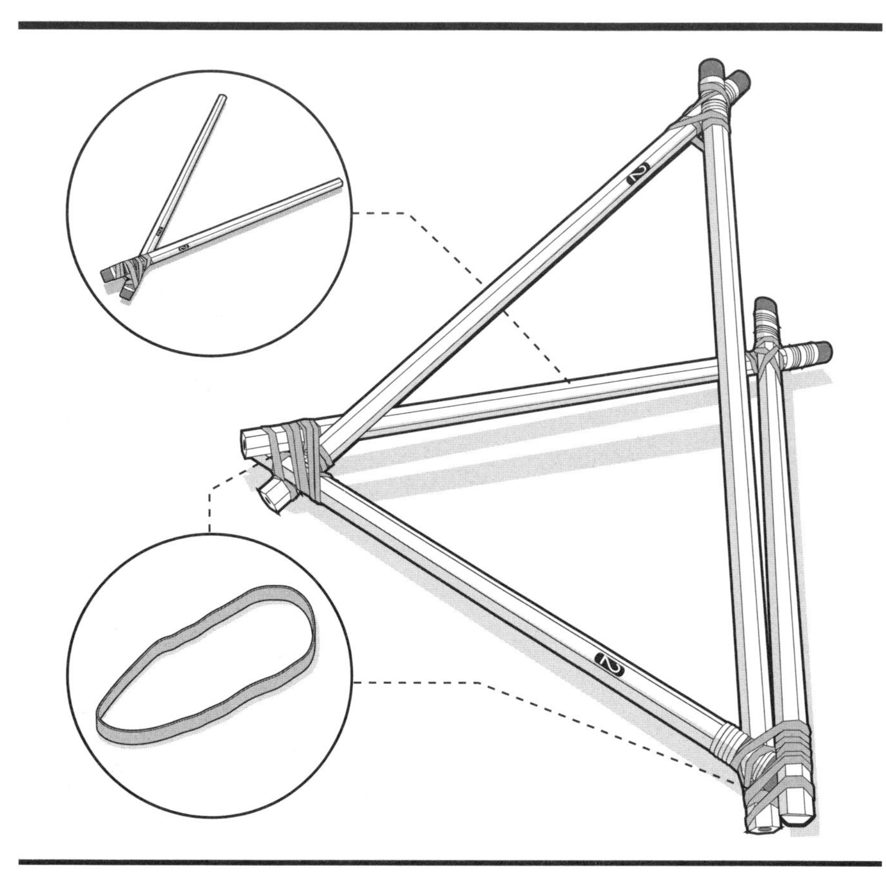

You will now combine the two assemblies from steps 1 and 2 to form a single frame. Take the two-pencil assembly and bend the pencils outward into a V shape. Rest the unattached ends of the two pencils onto the three-pencil triangular frame. Once in place, rubber band the two assemblies together to hold them in position.

## Step 4

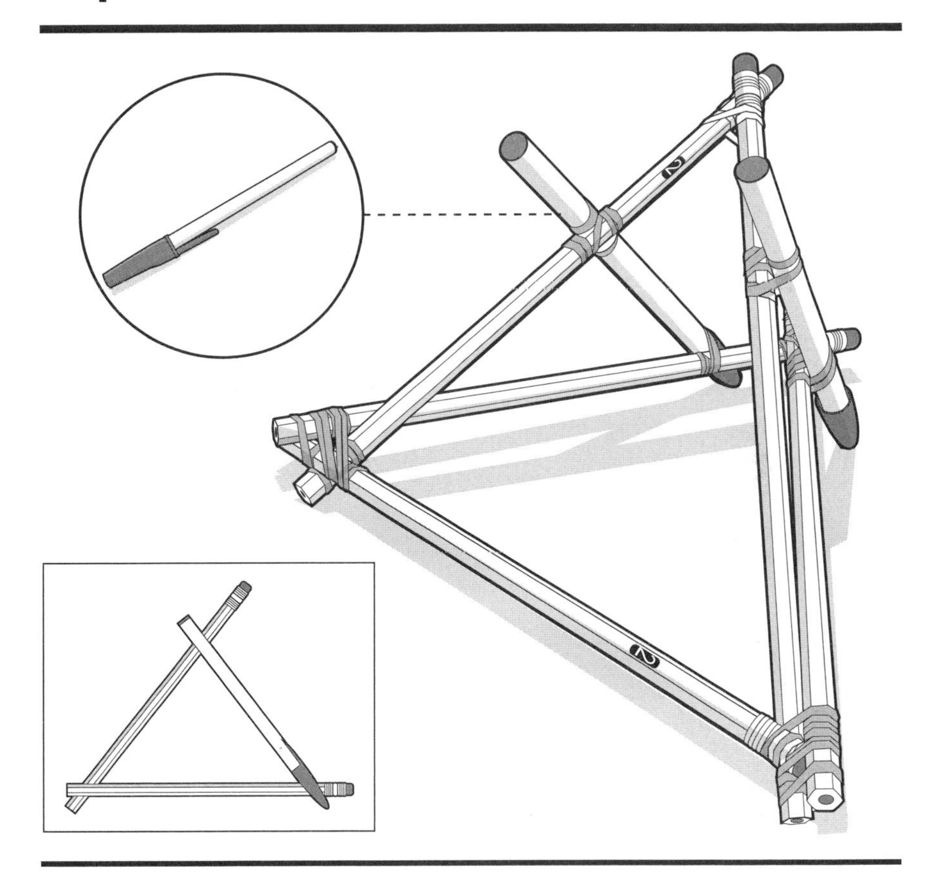

This step will complete your frame. Take two pens and place them on the sides of the triangle on the top halves of the pencils. Secure them in place with rubber bands.

Next, position the ends of the pens toward the back of the V-shaped pencil frame. Place the pens so they extend below the triangle base and become feet for the catapult. Once in place, use rubber bands to secure them in place.

The pens can be adjusted later to change the trajectory of the catapult.

## Step 5

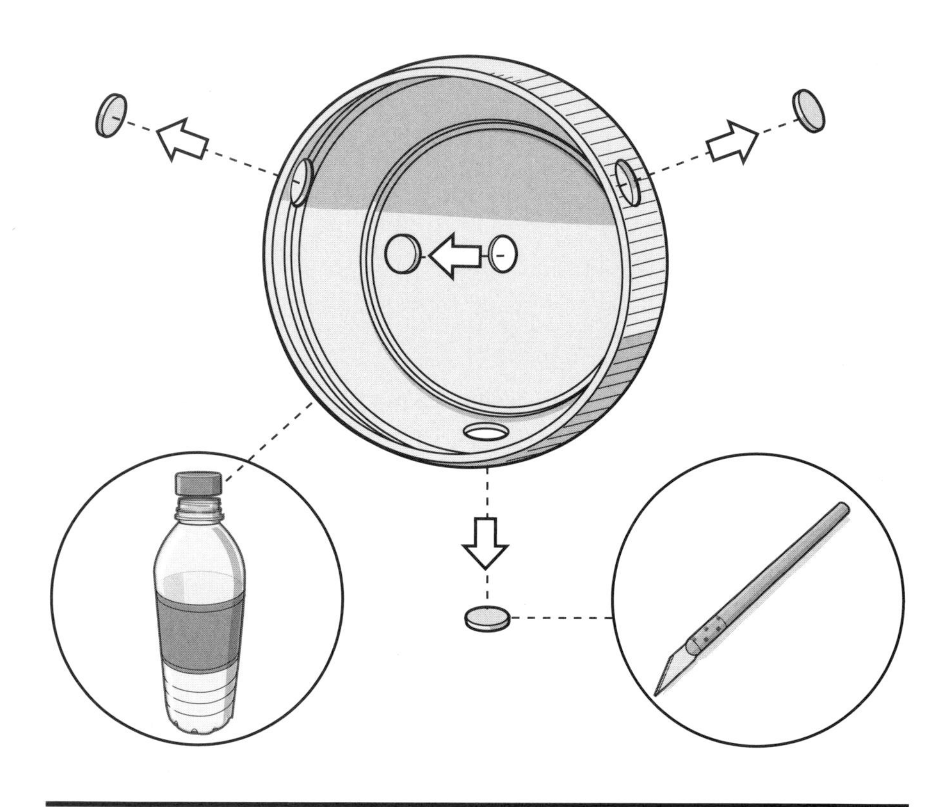

Now take a plastic cap from a soft drink bottle. Use a hobby knife to cut four small holes into the cap. The first hole will be dead center in the top. This hole should be approximately the same width as the rubber band you will use in step 6. The remaining three holes should be equally spaced around the side of the bottle cap, as shown.

Avoid getting too close to the edge of the bottle cap when cutting your holes. Insufficient material may give way when operating your catapult, resulting in malfunctions.

## Step 6

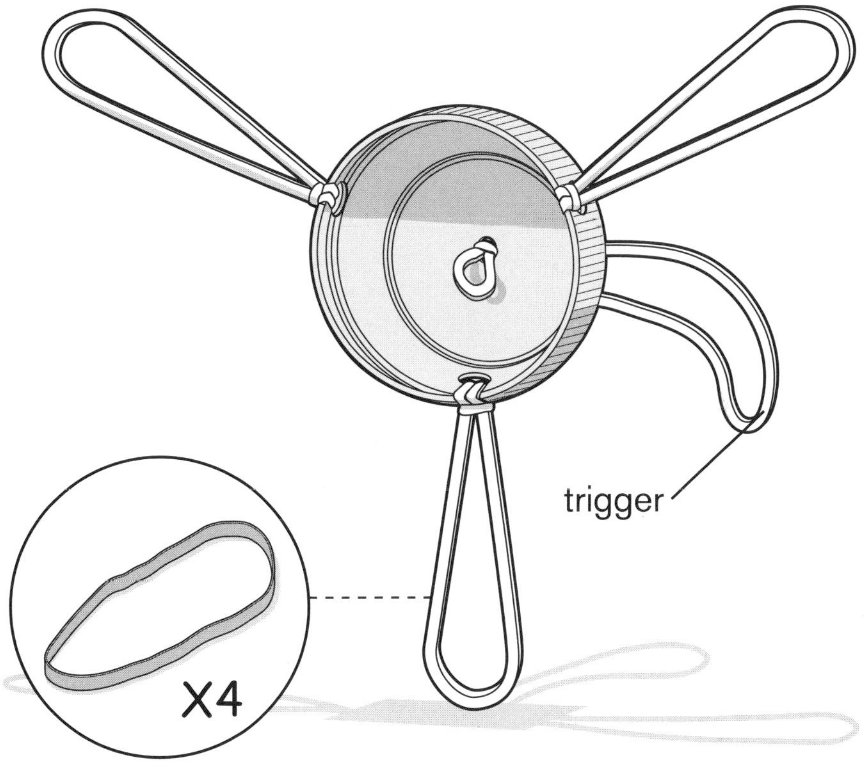

Loop one rubber band through each side hole and then over the cap and back through itself to form a knot. Each rubber band will then form a loop coming from the bottle cap.

The rubber band in the center should be pulled through the hole and then tied into a small knot on the inside of the cap. The knot must be considerably larger than the hole in the center of the bottle cap to prevent the rubber band from being yanked out.

# Step 7

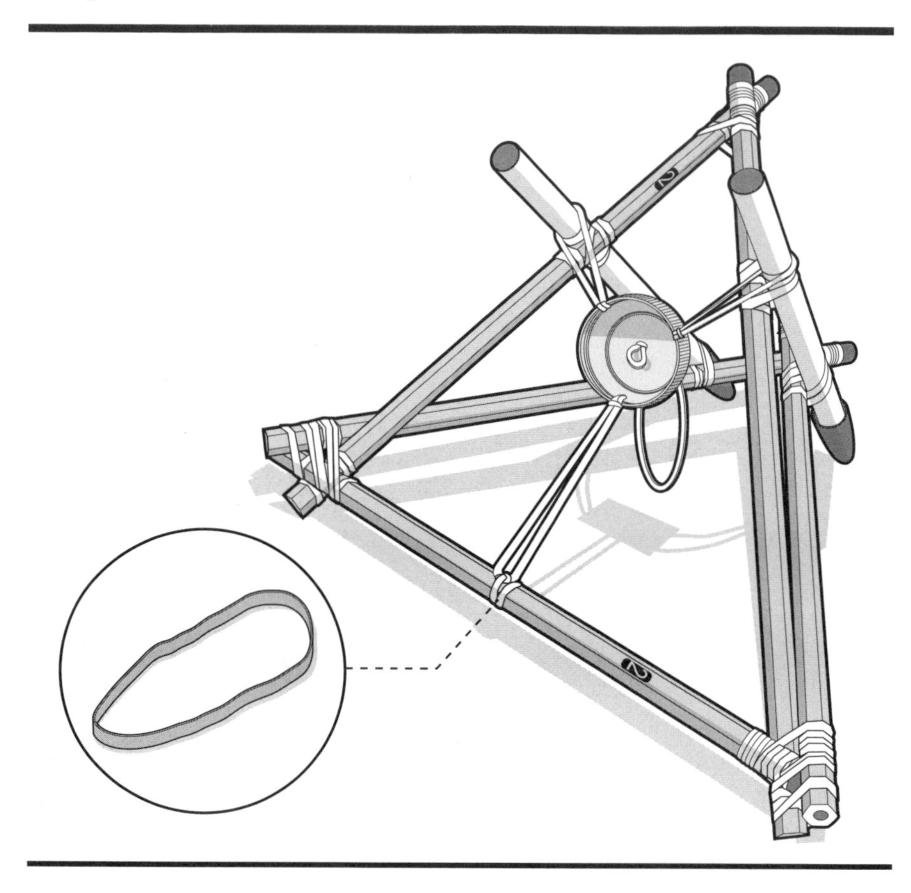

It is now time to install your bottle cap assembly onto its Viking Catapult frame.

Two of the rubber bands attached to the side of the bottle cap should be looped over the pen ends. The bottle cap should be positioned so that these two rubber bands are coming from the upper part of the cap; the third rubber band should hang straight down. Depending on the length of these rubber bands, you may want to wrap them a few times to tighten the connection. Pull the third rubber band, hanging off the bottom of the bottle cap, down toward the lower pencil and fasten it with a rubber band. That rubber band should be centered on the base pencil.

Now load your bottle cap with ammunition, pull back the centered rubber band attached to the back of the bottle cap, and release. If needed, adjust the pens to create the desired launch angle.

# PAPER-CLIP TREBUCHET

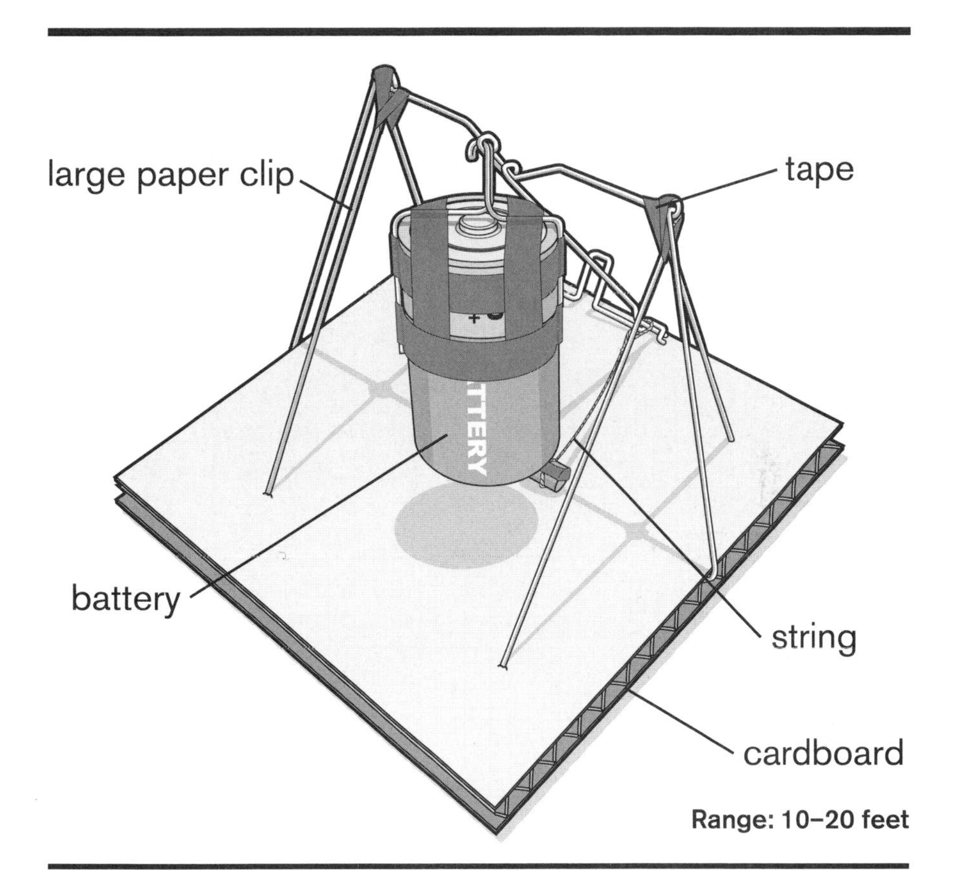


**Range: 10–20 feet**

The counterweight trebuchet is an accurate siege weapon dating back to the Middle Ages. It was designed to launch projectiles hundreds of feet, over or into enemy fortifications. Like its big brother, this Mini-Weapon will have to be adjusted for accuracy. But once you've mastered its construction, you'll soon be building larger versions capable of tossing pianos and compact cars over your neighbor's house. That's probably why trebuchets are sometimes called "bad neighbors."

## Supplies

8 large paper clips
Cardboard
Masking or duct tape
1 D battery
String

## Tools

Safety glasses
Needle-nose pliers
Scissors
Pen

## Ammo

1+ pencil erasers

# Step 1

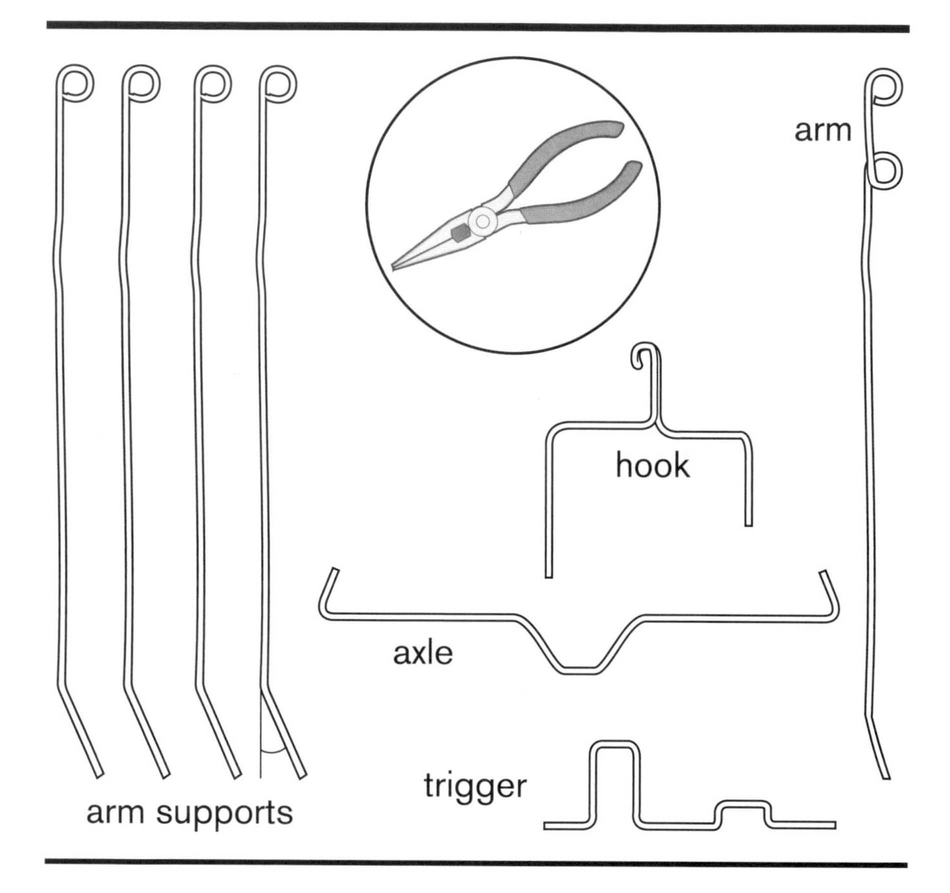


To begin, you must bend and cut eight large paper clips into various shapes and sizes. Use the illustration above for visual reference when manipulating the clips. Some additional bending and modification will take place as you construct the trebuchet.

First, straighten out all the large paper clips, then use the illustration above as an overlay as you rebend the straightened clips. To create the four arm supports, use a pair of needle-nose pliers to add one loop to the end of four clips. The opposite end of each arm support should have a slight angle bent into it.

To create the arm, bend the straightened clip in the same way as the arm supports, but add one additional loop under the end loop.

The hook should be bent to a width that fits around the D battery. Use the reference to bend the axle and trigger into their shapes; these paper clips may need to be cut down in size to match the overlay.

## Step 2

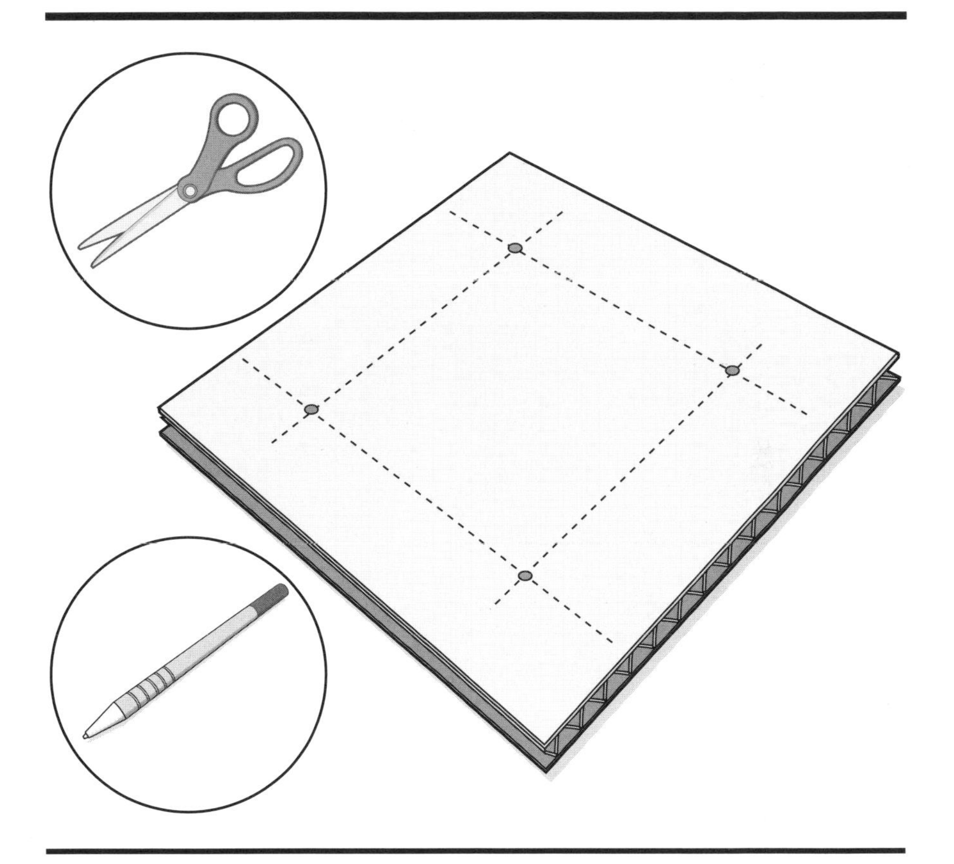

Bring on the base! Cut a piece of corrugated cardboard to be approximately 5½ inches by 5½ inches. Corrugated fiberboard from a shipping box is ideal because of the material's depth, perfect to burrow in the paper clip supports. If fiberboard is not available, thin cardboard can work as well, but some additional taping will be needed.

Now, with a pen, draw out a box that is approximately 3 inches wide by 3½ inches long, more or less centered on the cardboard surface. At the four corners of the box, draw circles to mark your four support holes.

## Step 3

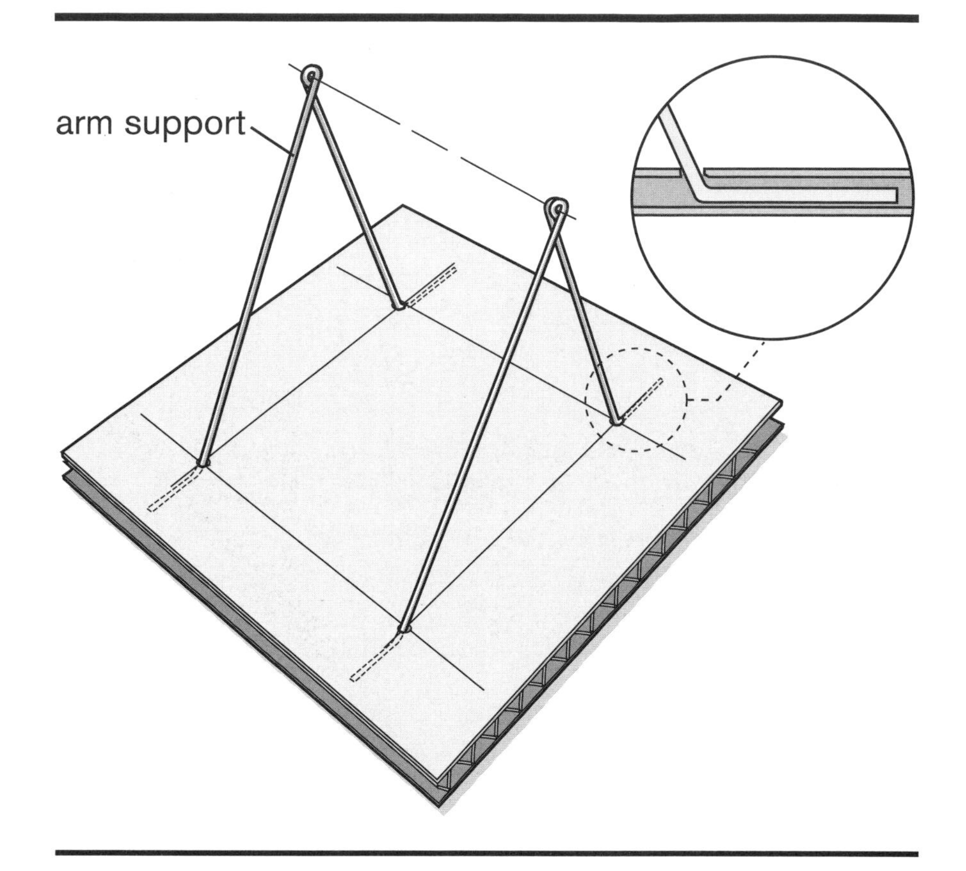


Now it's time for some support. Push the angled end of each support into the cardboard at the corners you marked on your box, facing out as shown. With the four support bars in place, line up the loop ends so they touch one another in pairs.

If you are using thin cardboard for the base, pierce the surface and tape the arm supports to the underside of the cardboard. This tape will not only hold the supports in place but will also prevent them from scratching the tabletop.

## Step 4

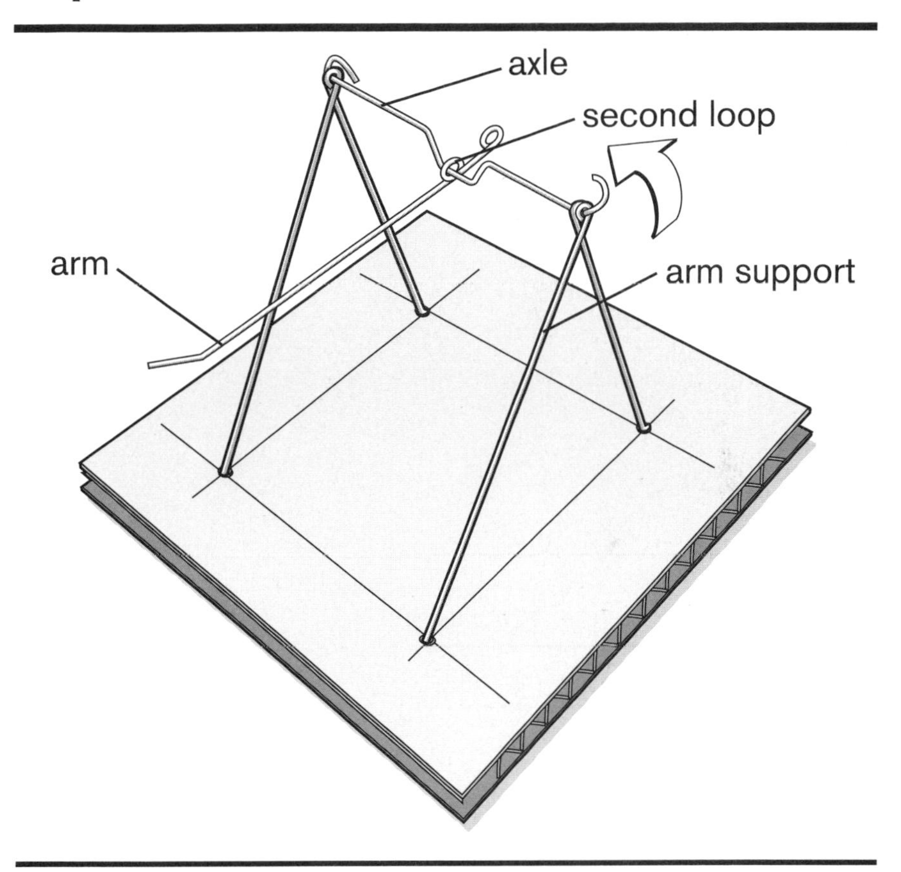


Now slide the axle through the second loop on the launching arm.

Once the arm is attached to the axle, slide the axle ends into the arm-support loops. Once in place, with the axle notch pointing down, bend the ends of the axle arms up to hold everything in place. You may need to use the pliers for this. The axle shouldn't be movable, so if needed, use tape to add additional support to the axle. Your design should resemble a child's swing set. If you feel your frame is still unstable, add two more paper clips for side supports, as seen on the final illustration (page 159).

## Step 5

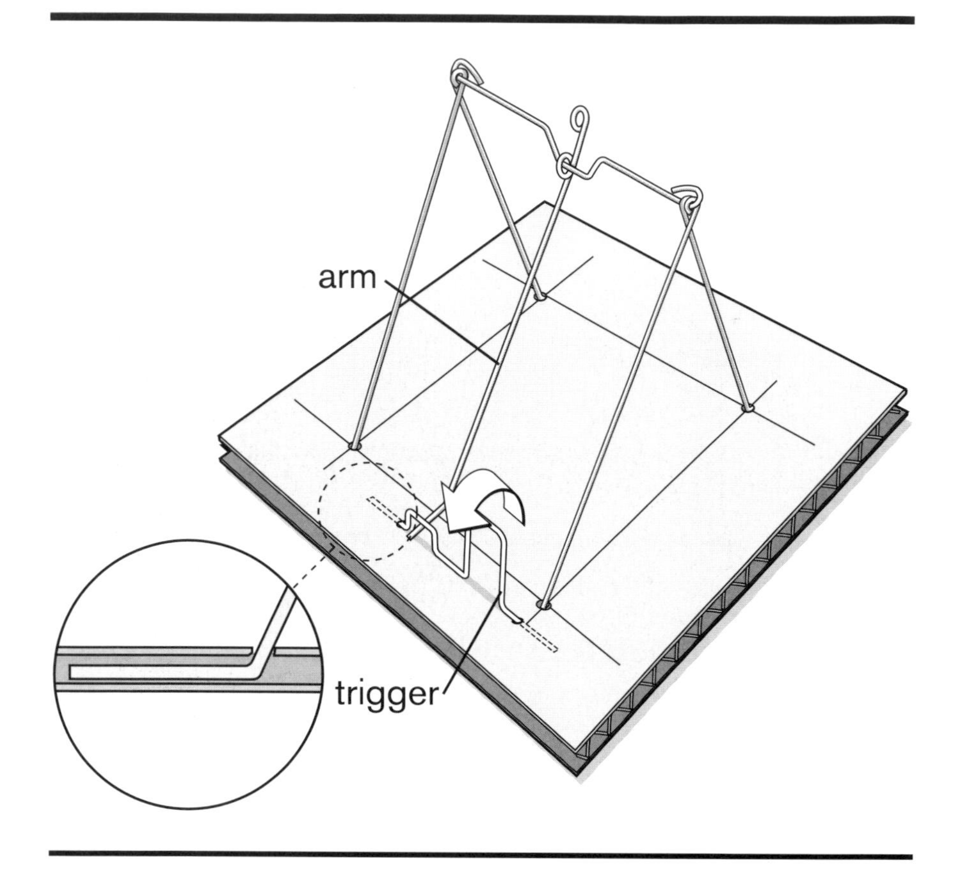


Now position the arm forward so that it's resting on the cardboard surface. This will determine the location of your trigger. You will be installing the trigger much like the supports, by running its ends under the cardboard. Poke the ends of the modified clip into the corrugated cardboard so that the small trigger notch aligns with the arm tip, as shown. The small notch in the bar holds the trebuchet arm, while the large notch is for your finger.

The trigger works when you flip the finger tap down, releasing the arm. Test your mechanism clearance before you install the battery weight.

## Step 6

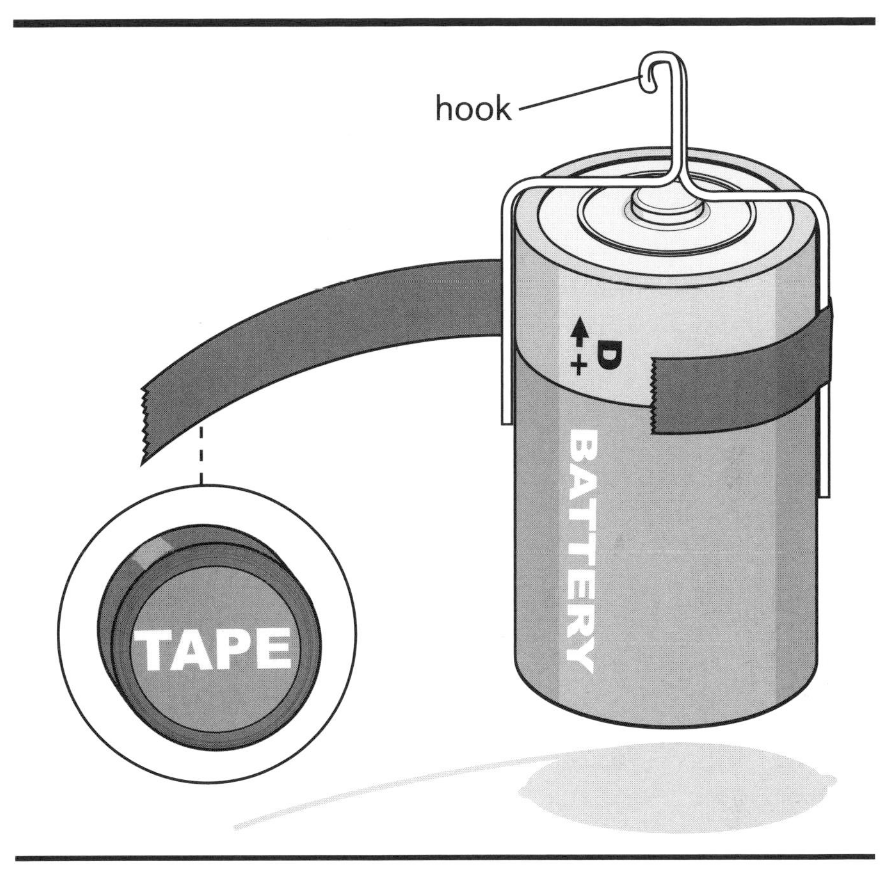


Bend the custom hook around the battery so it fits snugly around the housing. The hook should be located above the center point of the battery. Once correctly bent, securely tape the battery and hook together.

If a D battery is unavailable, three or four smaller batteries or a roll of coins can be taped together to create the counterweight.

# Step 7

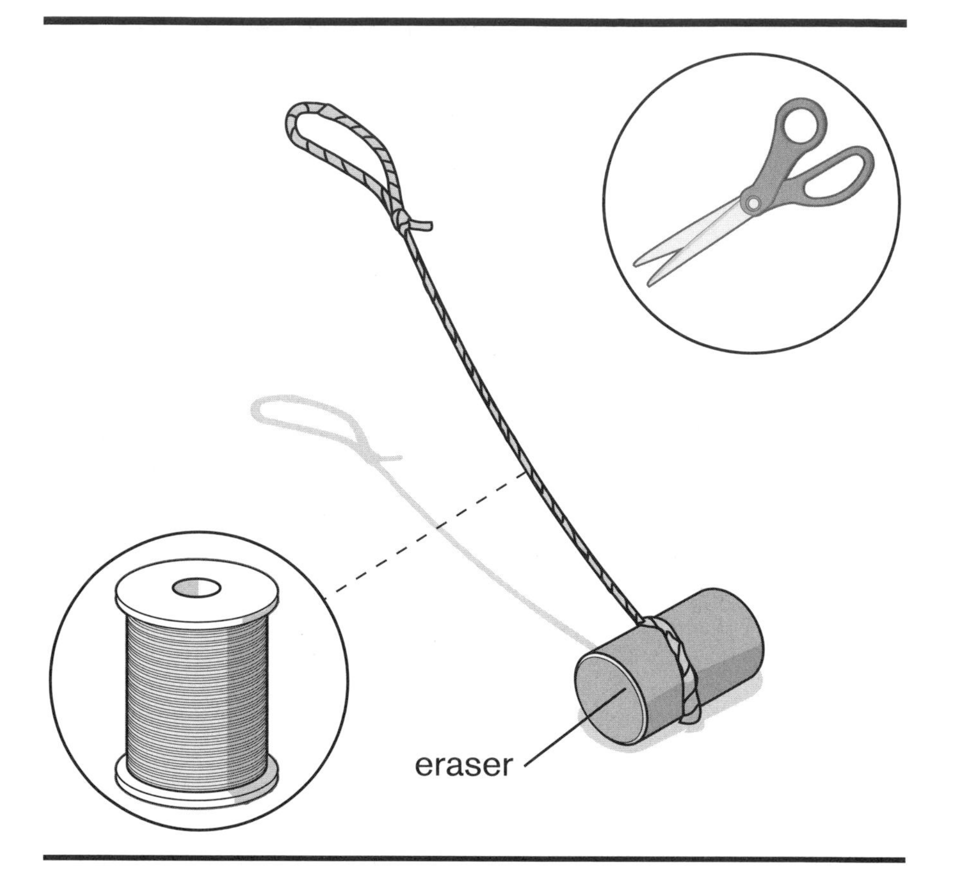


Now it's time to construct your ammunition. Because these things go flying, it may be worthwhile to make several before launching them.

Start with a 5-inch piece of heavy thread or kite string. Tie one end into a loop, then tie the opposite end around an eraser pulled from a pencil. If the knot doesn't look like it's going to hold, add some tape or glue.

## Step 8

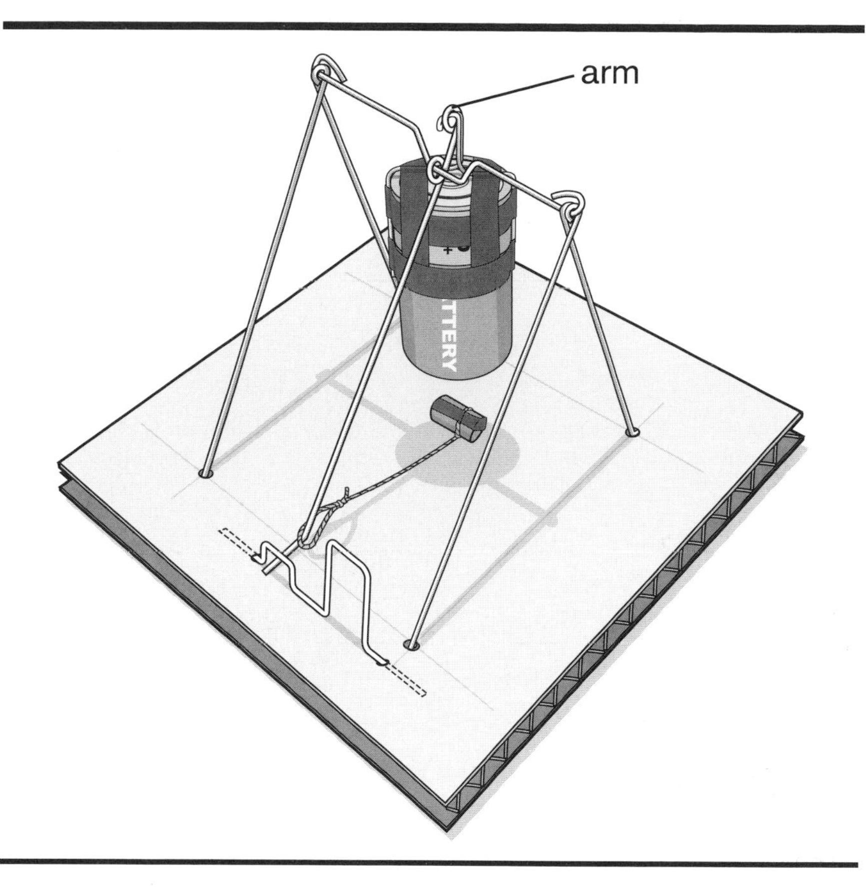


Your Paper-Clip Trebuchet is almost complete. Take the counterweight battery and hook it onto the last loop available on the launch arm, as shown. If you let it go, the weight should spring the arm upward. The battery should not touch the cardboard surface when the arm flings up, so if this happens, shorten the battery hook.

Next, place your ammunition loop over the opposite end of the launch arm. Place the string and eraser beneath the framework so they don't interfere with the trigger mechanism.

Now set the trigger so that it holds the launch arm down. When you're set to fire the trebuchet, put on your safety glasses, pull back the finger tab, and watch with amazement. Adjust as necessary for desired results.

# 5

# COMBUSTION SHOOTERS

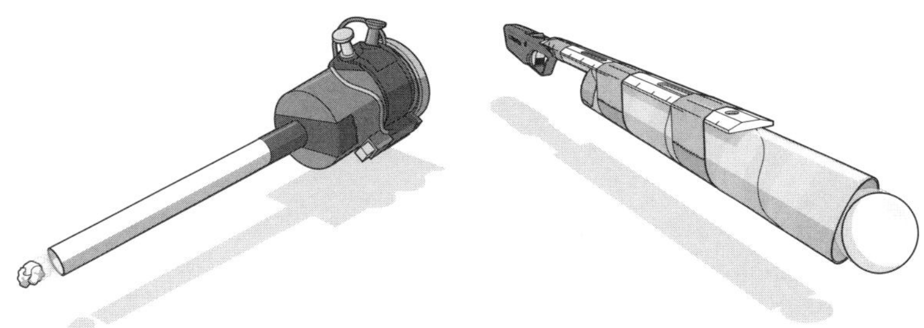

# AIRSOFT PEN POPPER

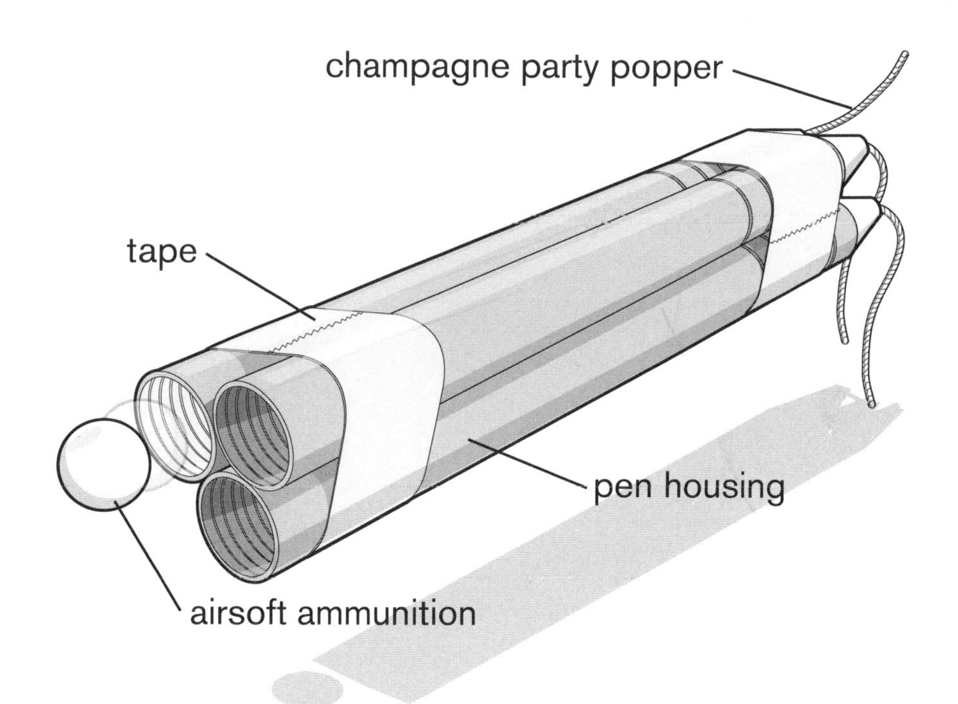


**Range: 20–40 feet**

It's party time! This multishooter packs a one-two-three punch against any target. With impressive firing velocity and a pocket-sized frame, it's the perfect companion for a well-equipped outdoorsman.

## Supplies

3 champagne party poppers
3 pens
Transparent or masking tape

## Tools

Safety glasses
Scissors

## Ammo

3+ airsoft BBs

# Step 1

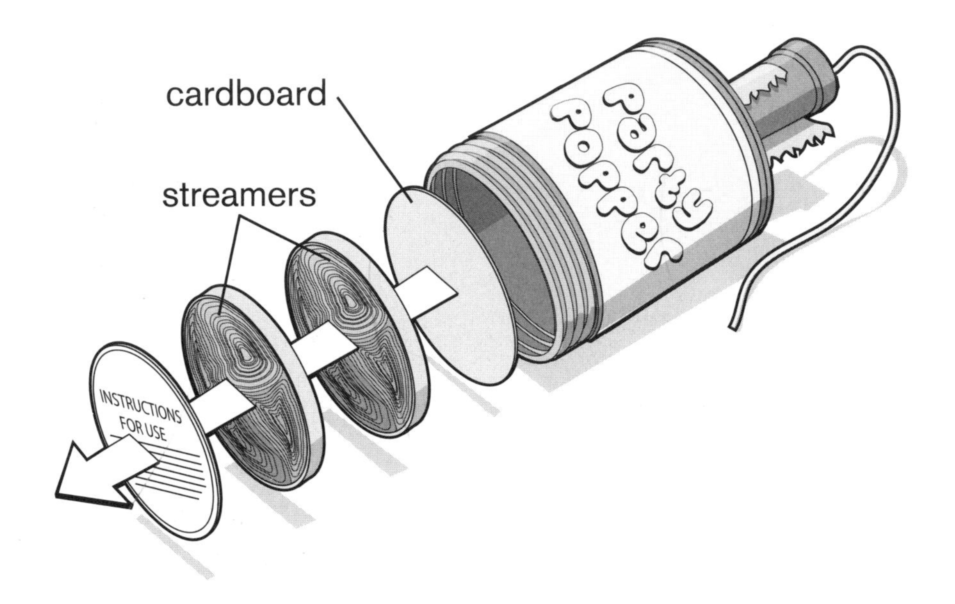


To start, you'll need party poppers that have a pull-string trigger. Also known as champagne party poppers, these loud products are designed to fill a room with colorful streamers. Because they are not considered fireworks, you can buy them at most major retailers that carry party supplies. Use your finger or the end of a writing utensil to remove the streamers and cardboard. This will make it easier to locate the explosive charge in the next step.

## Step 2

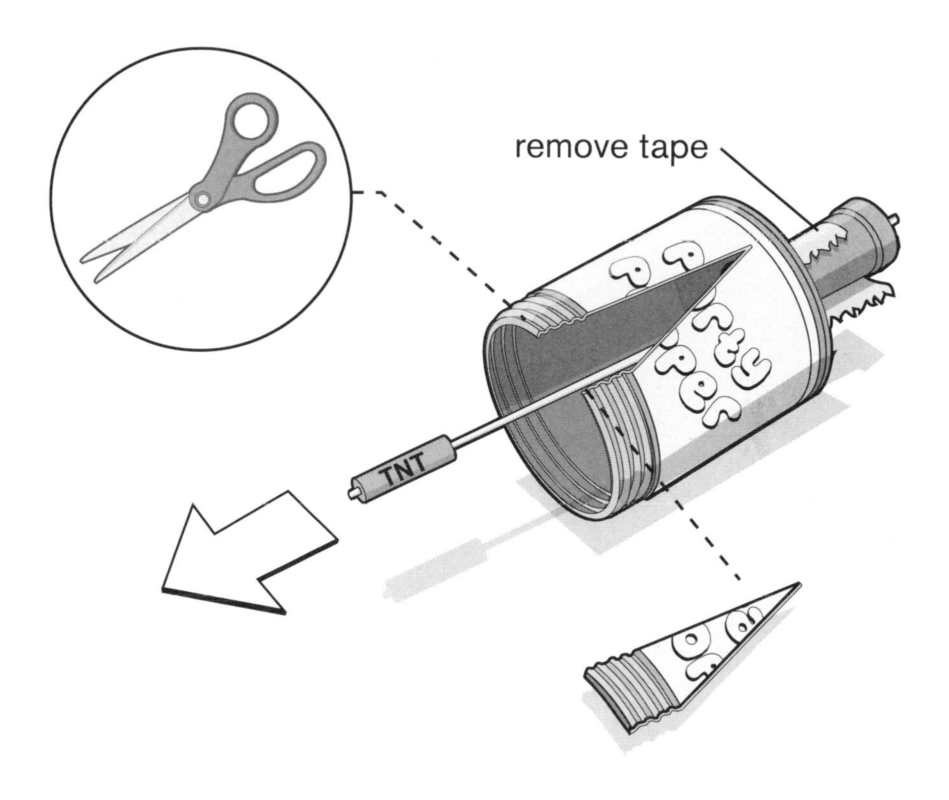


Now it's time to pull out the popper's explosive innards. Remember to put on your safety glasses. Then, remove the tape or decorative foil around the neck.

Using scissors, cut a section out of the side of the party popper. This will provide room for your fingers to grab the explosive charge. Once you have it, slowly slide it out with the string still attached. If it feels like it's not coming, ***don't yank it***—this might detonate the explosive charge. Once the charge is removed, you can discard the popper housing.

# Step 3

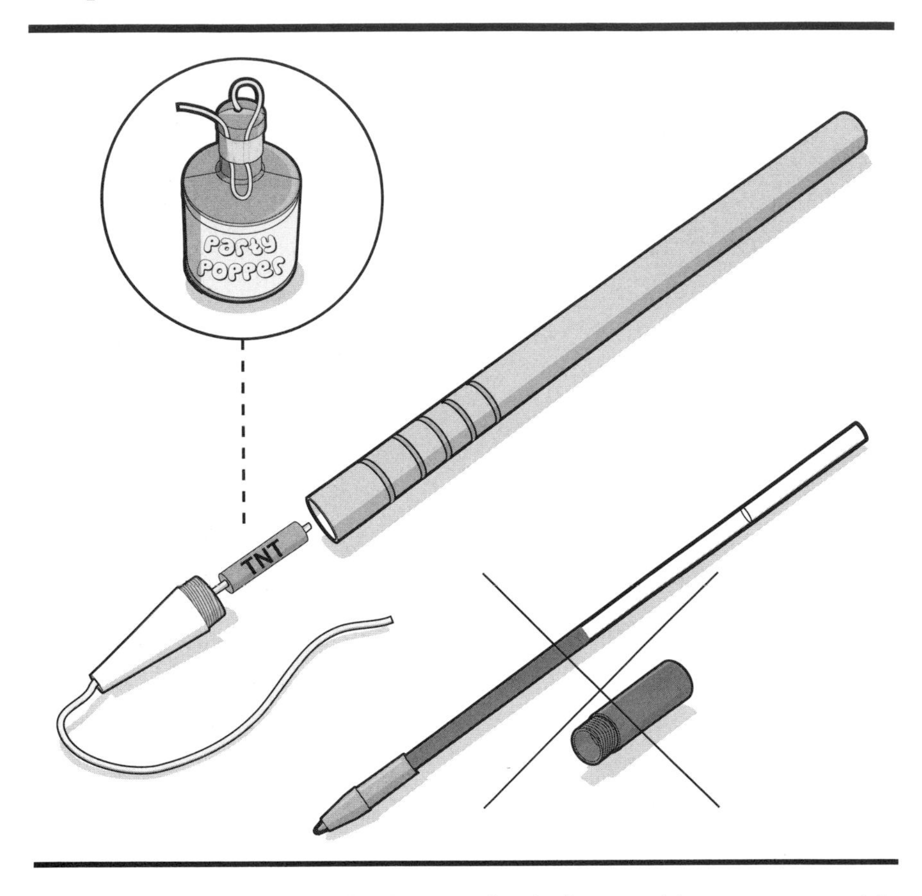


Your pen should be a ballpoint or roller-ball pen with an unscrewable metal tip. These are generally inexpensive, and you likely have a few lying around.

Disassemble the pen and remove its ink cartridge, metal tip, and plastic end. Discard the ink cartridge and plastic end.

Now take your party-popper charge and push the string through the metal pen tip. Gently pull it all the way through until the charge is nestled into the metal cone. Once it's in place, screw the tip back into the pen.

## Step 4

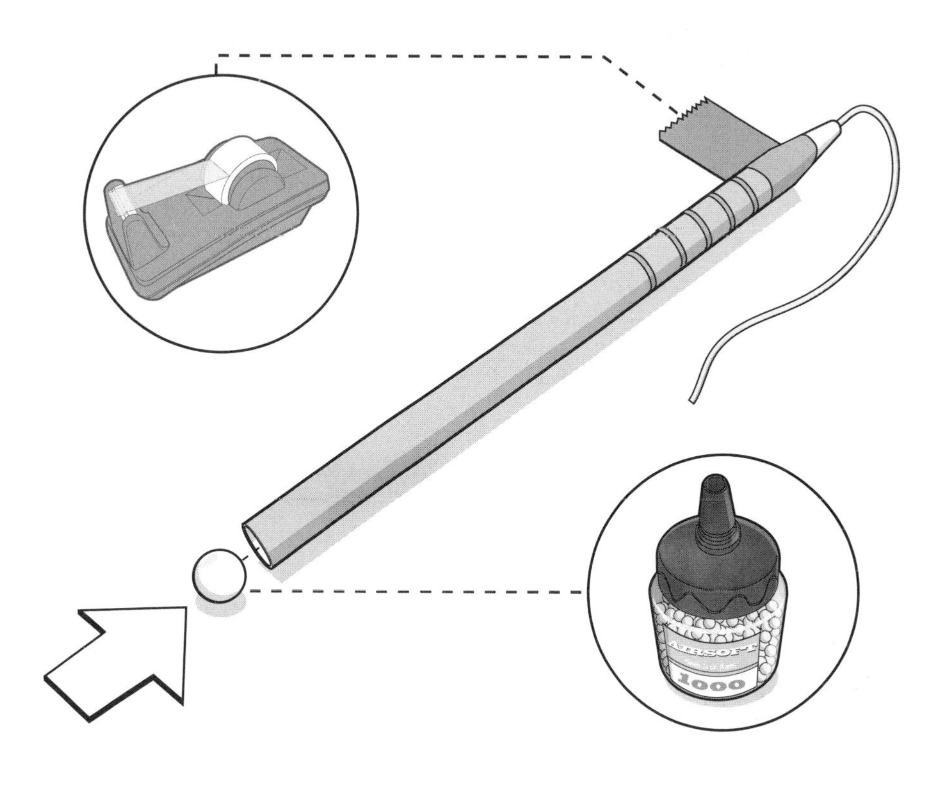


Next, tape the metal tip onto the pen as an extra precaution. Depending on the manufacturer and materials, there is always the chance for the unexpected to occur.

Load your .24 caliber (6 mm) airsoft ammunition into the muzzle end of your barrel. These BBs are very inexpensive and usually fit perfectly into pen housings. To fire the popper, simply aim and pull the popper string.

## Step 5

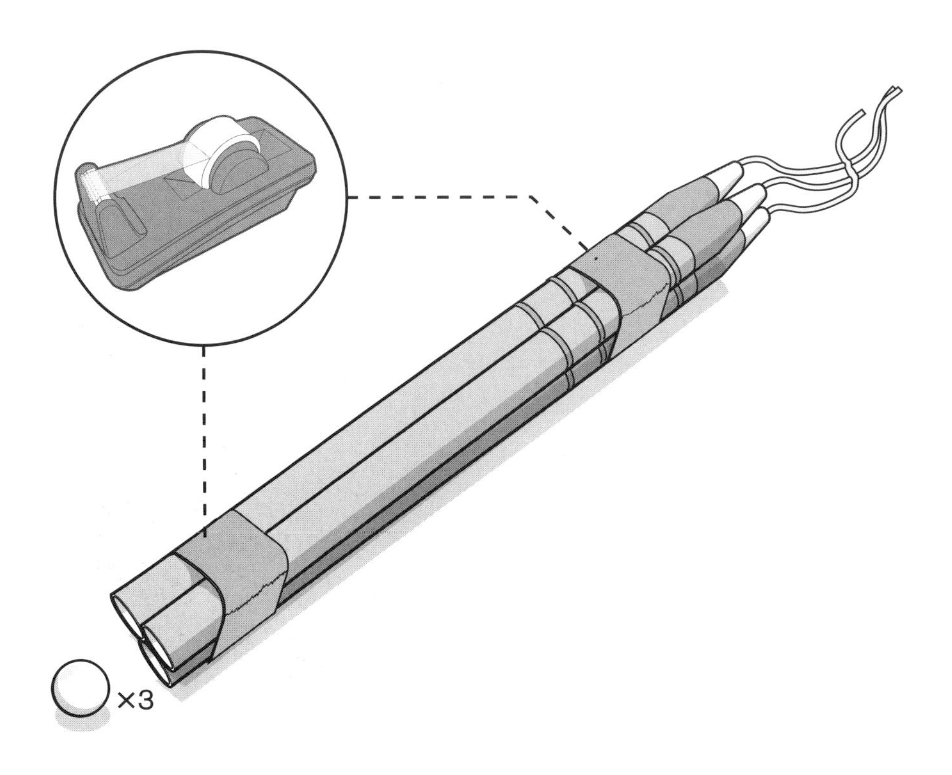


If you want to fire multiple shots in a short period of time, tape three pen housings together, each preloaded with an explosive charge. Pull the strings individually to give yourself multiple shot opportunities.

You may substitute ammunition and use a good old-fashioned spitball or the original streamers. Just use one of the ink cartridges to ram the ammunition down into the pen barrel prior to firing. These are safer alternatives than BBs, but you will not achieve the same firing distance.

Never aim your Airsoft Pen Popper at a human or animal! A BB ricochet is probable, so always wear safety glasses when firing. And ***never look down the barrel***, even with safety glasses on.

# MATCH ROCKETS

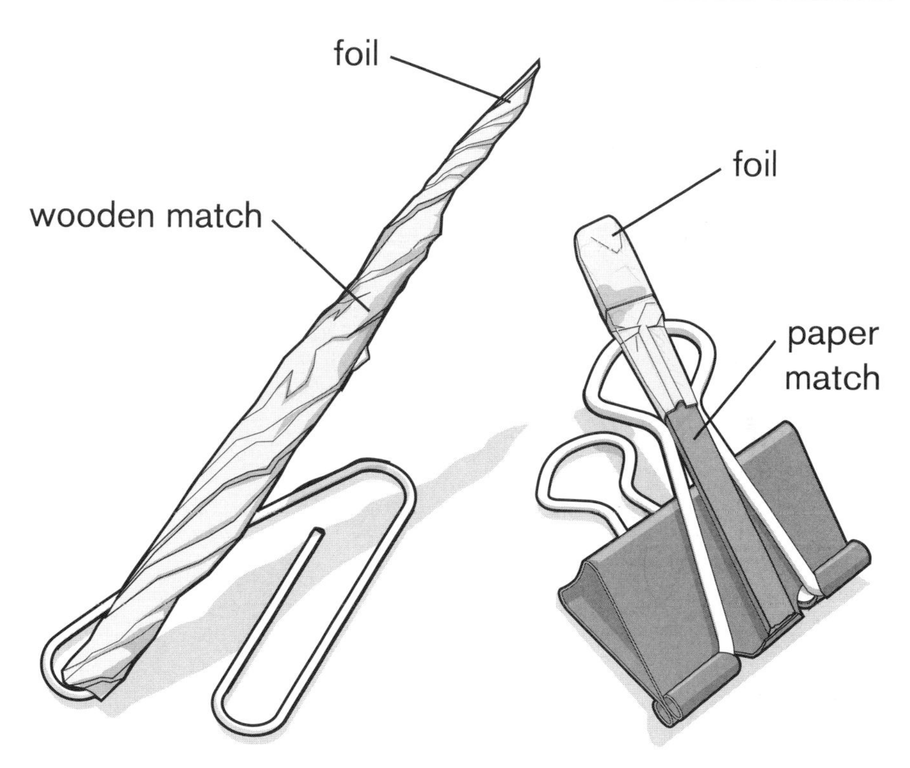


**Range: 20–40 feet**

Match Rockets are a very inexpensive way to learn about basic rocketry. There are a few basic types of Match Rockets, which can be made with paper matches or wooden matches. They are unpredictable and can fly in any direction, making it interesting to see where they might land. ***Eye protection and a safe firing range are musts*** when experimenting with and exploring these minijets. Match rocketry is not an exact science—misfires and modifications will be needed to find the perfect balance.

## Supplies

Aluminum foil
1 needle or pin
1 medium binder clip (32 mm)
1 toothpick
1 large paper clip

## Tools

Safety glasses
Pocketknife

## Ammo

1+ paper matches
4+ wooden matches

# Paper Match: Step 1

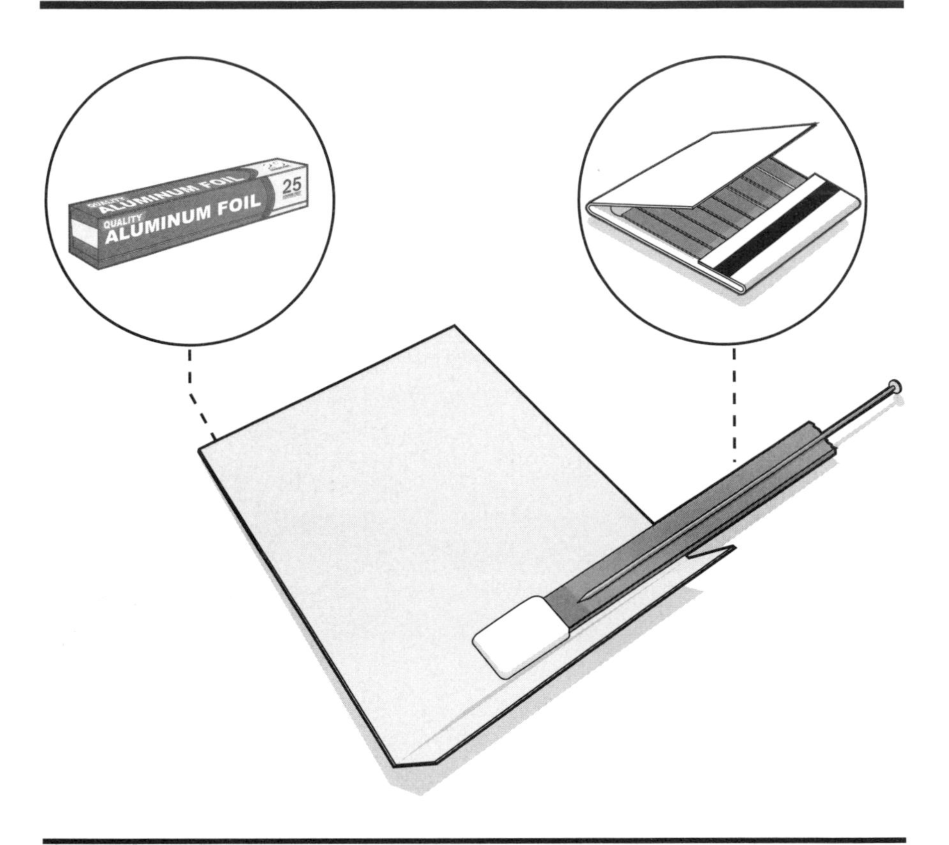


Start by building a Paper Match Rocket. Remove a single paper match from a matchbook. Place it onto a small piece of aluminum foil, no more than 1 inch by 1 inch. Then place a needle or pin onto the paper handle, as shown. This needle will form your exhaust port in the next step.

## Paper Match: Step 2

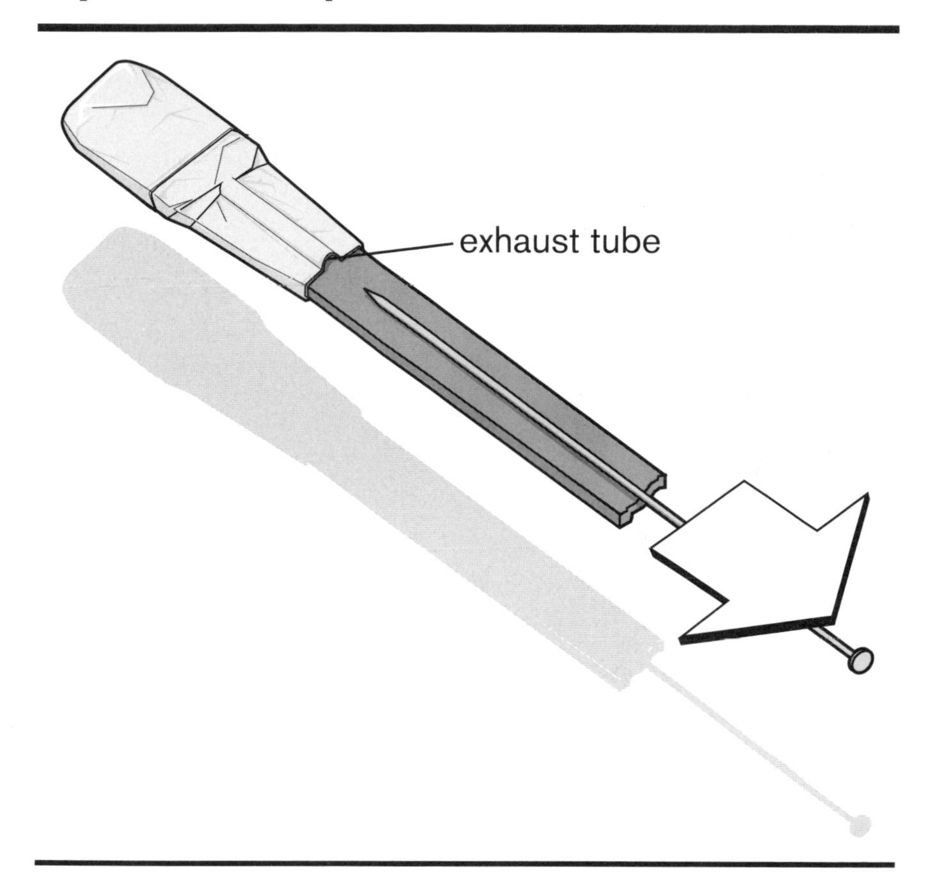


Roll the aluminum foil over the top half of the match. Make sure you tightly form the foil over the phosphorus head of the match and around the pin or needle protruding from it.

Once the match is wrapped, slowly slide the needle out of the aluminum wrapping without reshaping the foil. The pin has created an air channel that goes directly to the match head. This will be the rocket's exhaust port. Handle the rocket with care before launching to avoid crushing this channel. A crushed exhaust port will cause a blowout or misfire.

## Paper Match: Step 3

Put on your safety glasses. Place your Paper Match Rocket on a binder clip launcher, head up, pointed away from you. It's a good idea to protect the surface from which you're launching your rocket. A Paper Match Rocket is capable of burning surfaces and igniting flammable material, so careful outdoor use is a must.

Now hold a lit match beneath the rocket's head. Wait a few seconds for the flame's heat to penetrate the aluminum foil and ignite the paper match. Once lit, the burning head will create pressure and shoot gases out the exhaust port of your minijet. This force will propel your rocket into space.

If your rocket just lights up but doesn't fly, it's because the gases couldn't make it down the exhaust port. Rebuild and try again.

## Wooden Match: Step 1

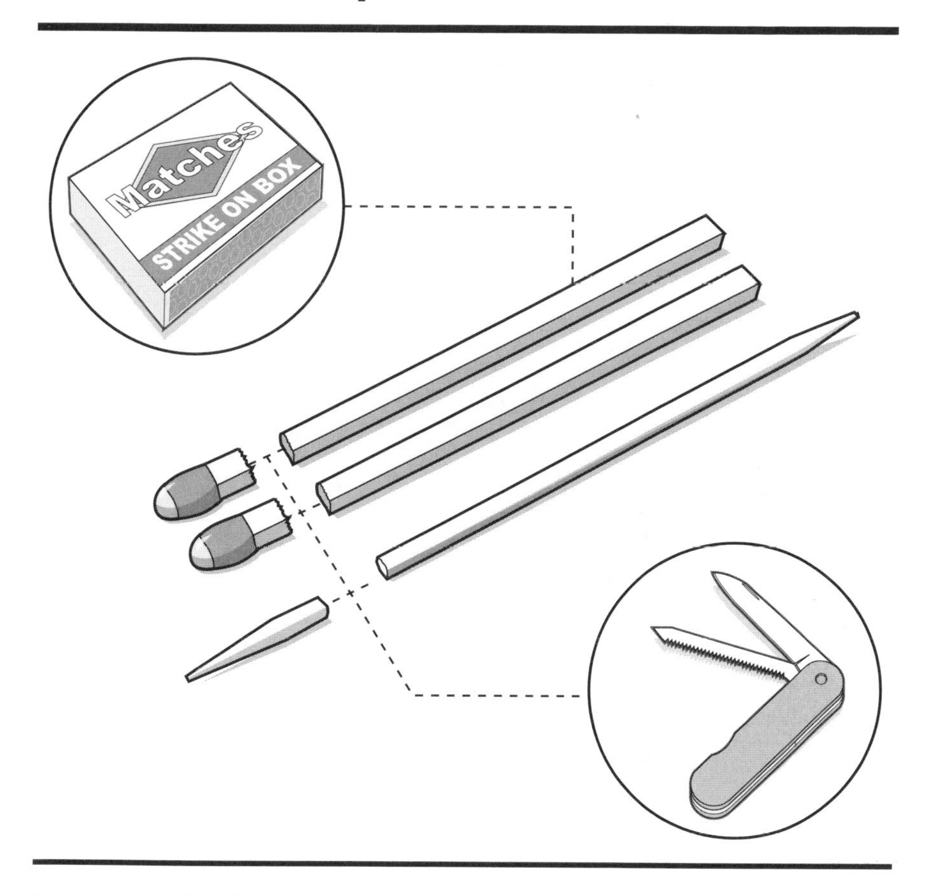


I personally find Paper Match Rockets the easiest type of match rocket to fire. However, I've also found that Wooden Match Rockets fly the greatest distance, which is why I've included designs for both. Wooden matches vary by manufacturer, as do their rocketry results. Diamond is one brand of match that does *not* work.

Now it's time to build a Wooden Match Rocket. Take two wooden matches and sever the phosphorous-coated head with a pocketknife. Keep the heads and discard the sticks.

Next, take a round wooden toothpick and remove one of the points from the end with a pocketknife. Discard the point and save the toothpick.

# Wooden Match: Step 2

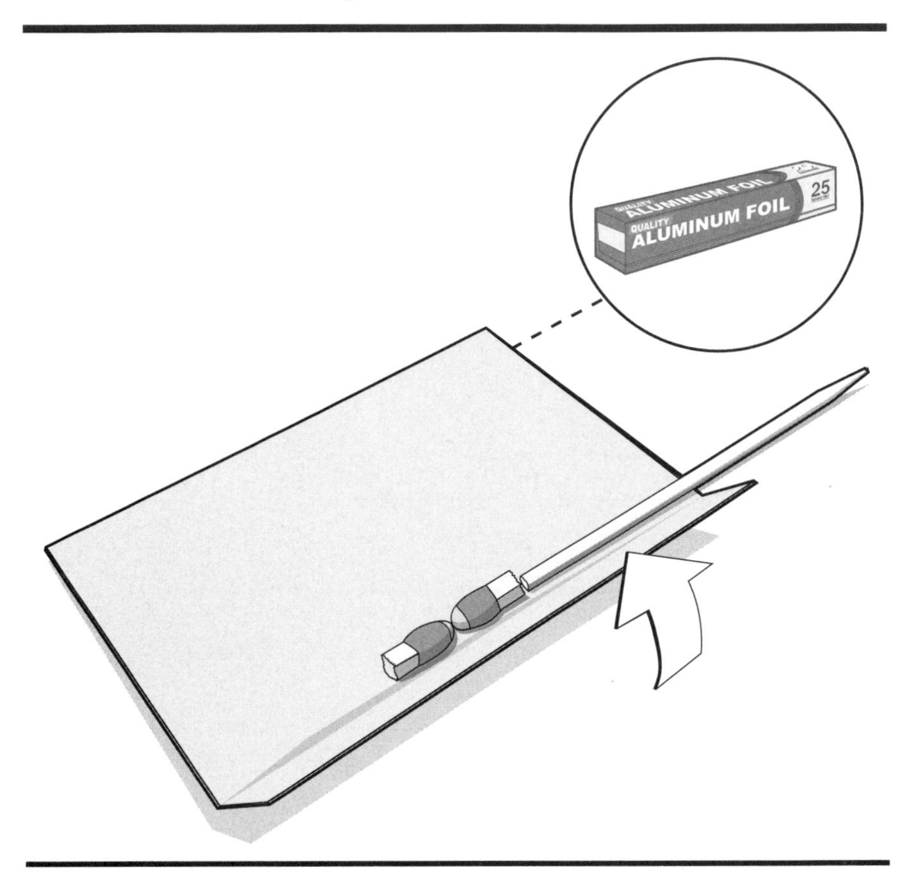


Cut a small square of aluminum foil with sides that are approximately the same length as a wooden match. You must have enough material to roll up the match heads several times. If the walls of your rocket are too thin, you will have a blowout problem and a fire on the launch pad. But on the flip side, if the walls are too thick, the added weight will decrease the distance your rocket travels.

Along one edge of the foil, make a straight fold. This crease will help you line up the materials. Place the two match heads facing each other on the inside of the fold, an approximately equal distance from the edges of the foil. Then place the cut, blunt end of the round toothpick behind one of the match heads, as shown.

## Wooden Match: Step 3

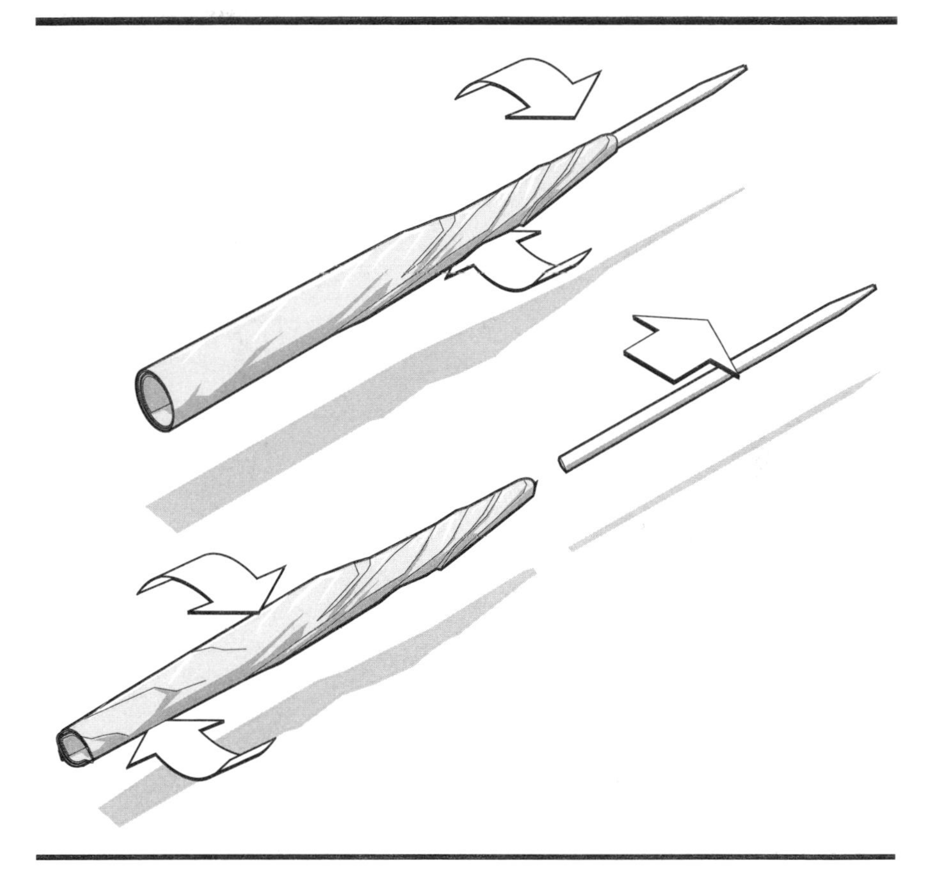

Without spilling the match heads, tightly roll the foil around them and the toothpick. Then roll the front of the aluminum rocket head to form a crude missile tip. Finally, slowly pull out the wooden toothpick. The toothpick will have created a cylindrical hole leading up to the match heads. Do not disturb or squeeze this end or the exhaust port.

Your Match Rocket is ready for the launch pad!

## Wooden Match: Step 4

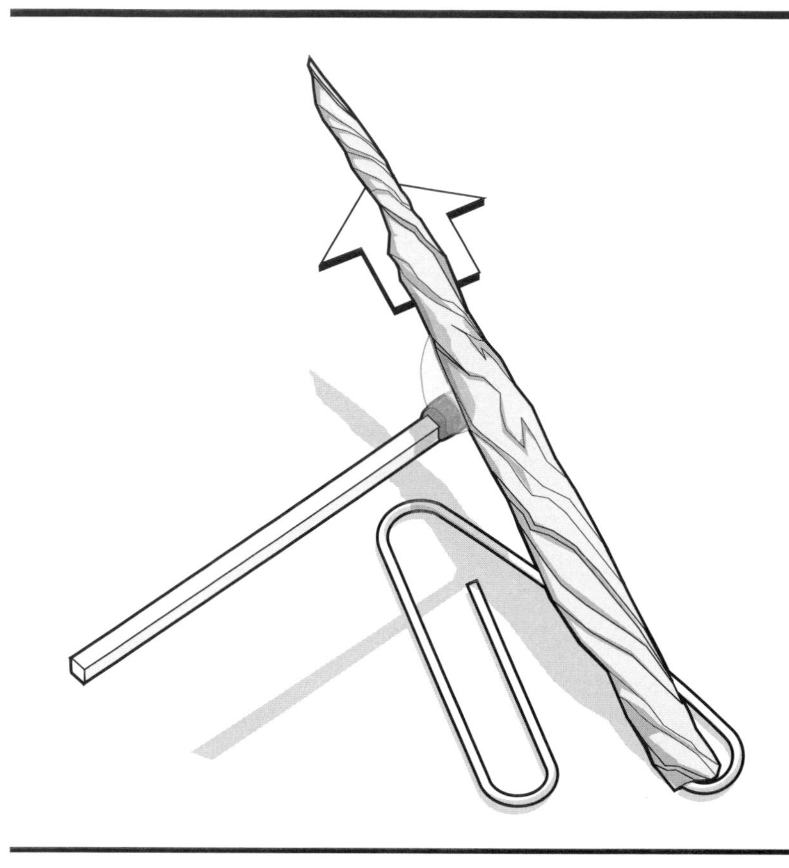

Make a launch pad for your Wooden Match Rocket by partially unbending a large paper clip to form the base, then pivoting one of the ends upward to form a guide for your rocket. Once your launch platform is bent into shape, slide the bottom of your rocket over the end of the paper clip launcher.

Put on your safety glasses. Hold a burning match under the covered match heads and wait for ignition. If the match heads burn through the foil wall, it means either that your rocket needs to be wrapped a few more times or your exhaust port was blocked.

Remember, you are shooting a flaming match, so use these rockets outdoors and remove all flammable materials from the launch zone.

# MINI SPUD-AND-SPIT

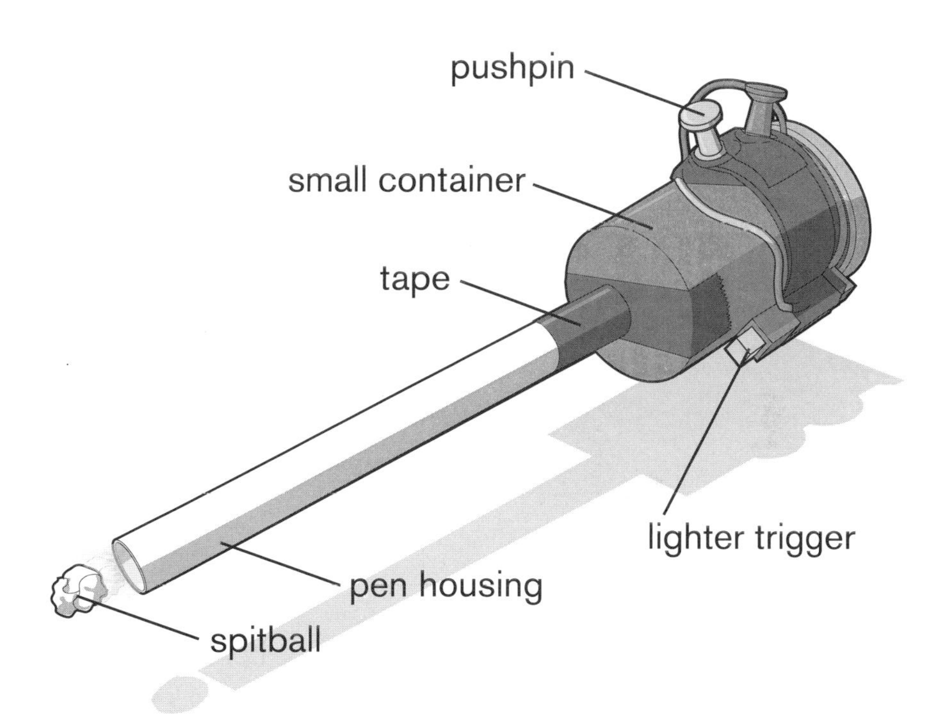


**Range: 20–40 feet**

This Mini Spud-and-Spit gun is the perfect solution for launching spitballs. Similar to a firearm or spud gun, it works by pushing the projectile out of the barrel using expanding gas.

## Supplies

1 barbecue lighter
2 pushpins
1 small container (with cap)
1 pen
Electrical tape
Hairspray

## Tools

Safety glasses
Screwdriver
Wire cutter
Hobby knife

## Ammo

Spitball

# Step 1

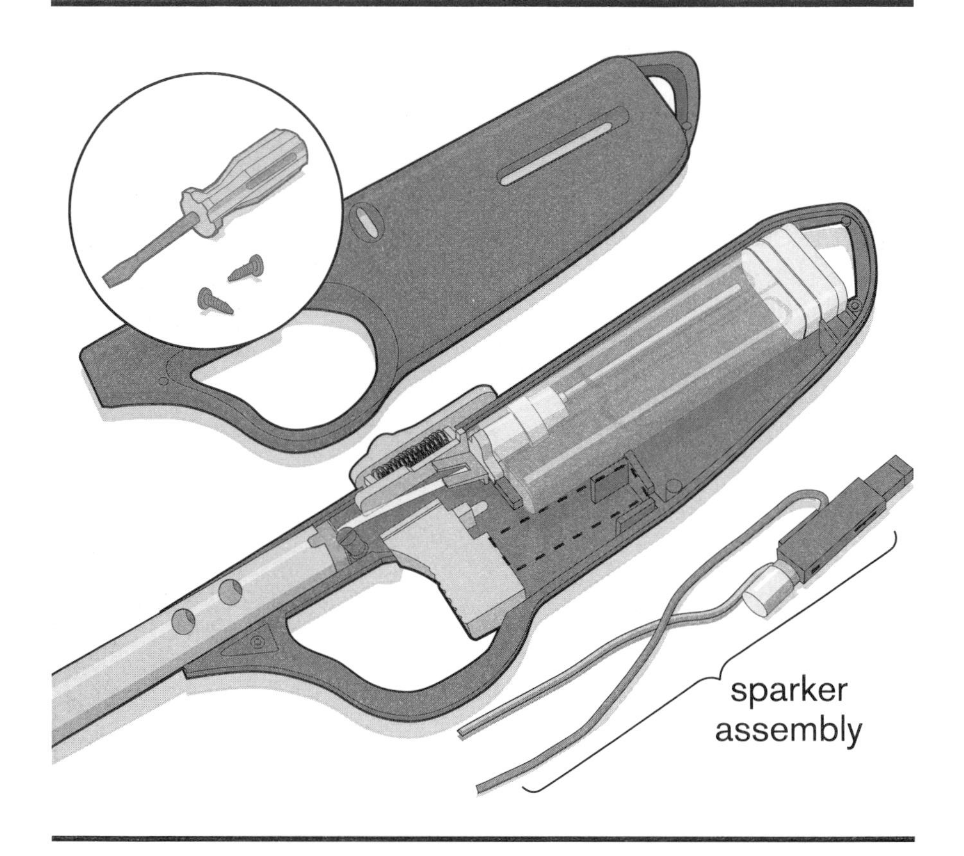


Focus first on the ignition source. Using a screwdriver, disassemble your barbecue lighter. Once you've opened the lighter, locate the sparker assembly behind the trigger. Remove the unit and wires, keeping them intact, but be careful not to shock yourself by pushing the button. (You will not need the rest of the lighter, so discard the other components, including the fluid.)

## Step 2

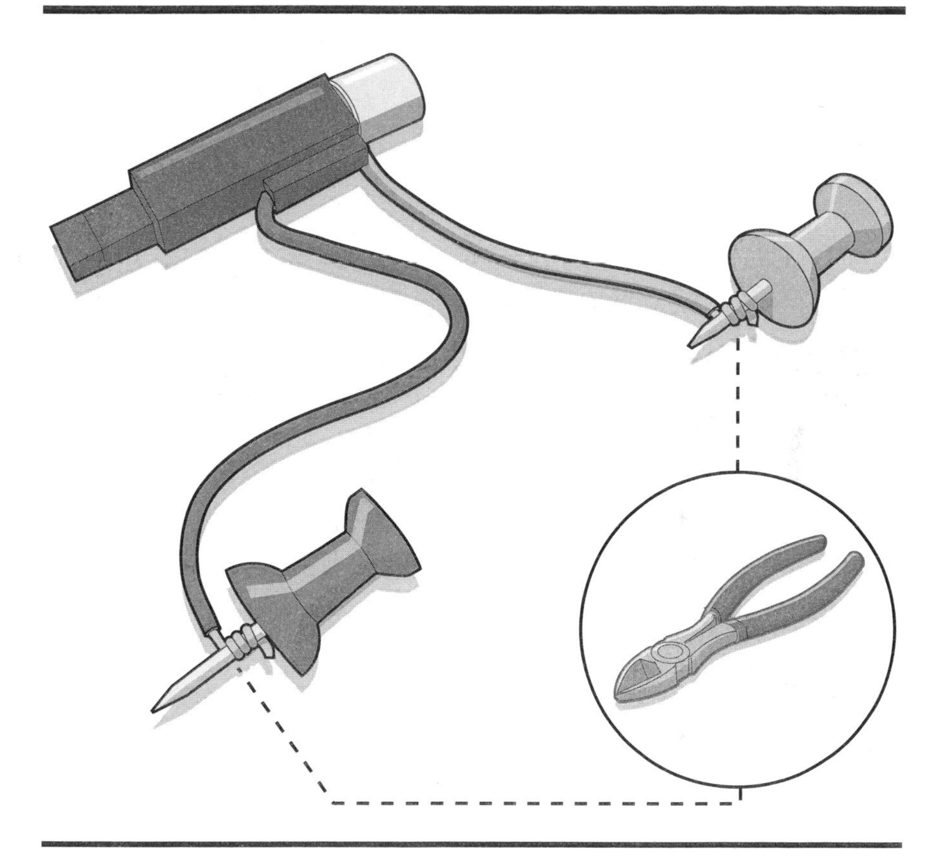

Using a wire cutter, trim the plastic insulation from the last inch of both wires. Be careful to avoid cutting the wires.

Tightly wrap each exposed wire end around the metal point of a pushpin. These pushpins will act as your electrodes in the Mini Spud-and-Spit's combustion chamber.

## Step 3

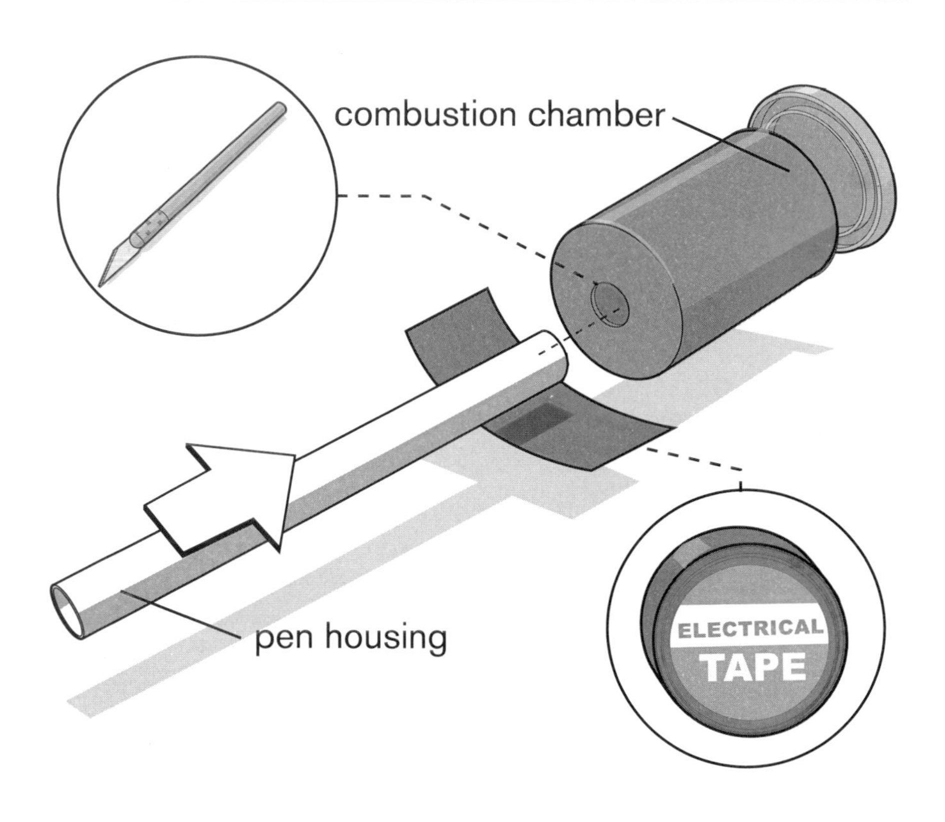


Your airtight combustion chamber must have a cap that can be easily removed and closed during operation. In this design, you will be using an old film roll container for the chamber.

Empty a pen housing of all its contents. Then, with a hobby knife, carefully cut out a hole in the bottom of your combustion chamber; the hole should have a diameter exactly the same size as a pen-housing barrel.

Once you have carved out a hole, slide the barrel into the combustion chamber and tape it in place. The barrel and combustion chamber should be airtight, so don't be stingy with the tape.

## Step 4

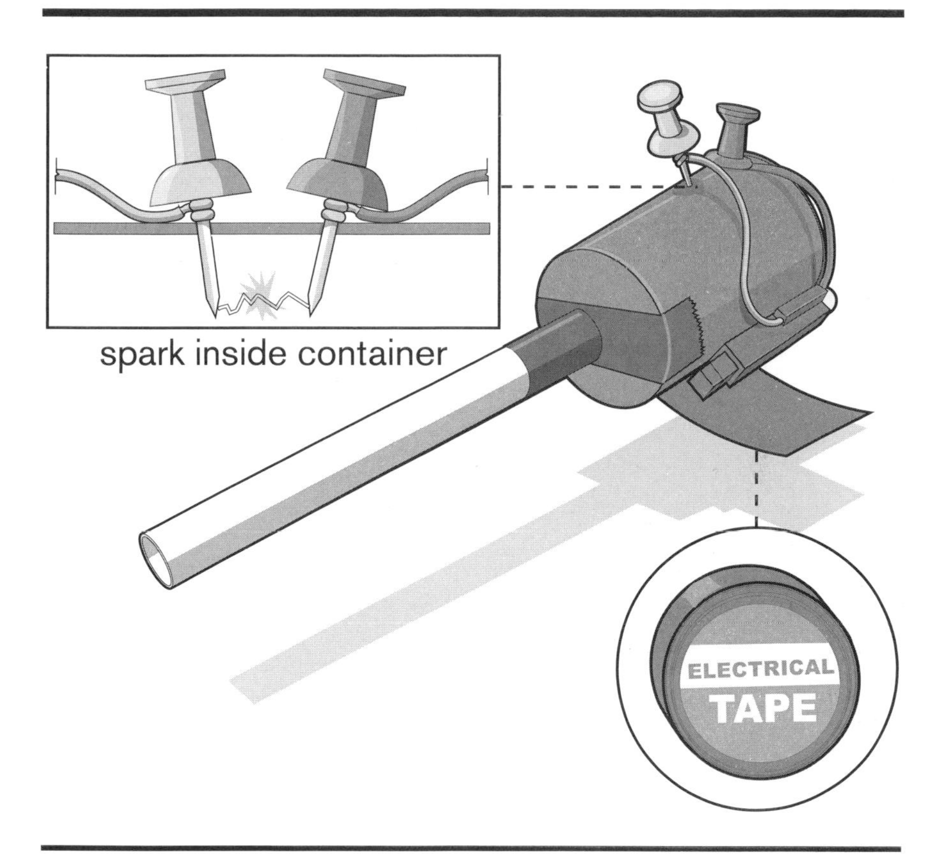


Now it's time to attach your ignition system. Stick the pushpins into the combustion chamber near the rear of the chamber roughly a quarter of an inch apart from one another. Then, using electrical tape, attach the trigger from the barbecue lighter to the side of the film canister. Before testing, wrap vigorously with electrical tape.

Once the trigger is taped on, test your ignition system. Push the small switch on the trigger and make sure the pushpins are sparking inside the chamber.

If your ignition system is not sparking, the metal ends of the pushpins may not be close enough or the electricity may be arcing outside of the canister, so rework the wires until you get it right.

# Step 5

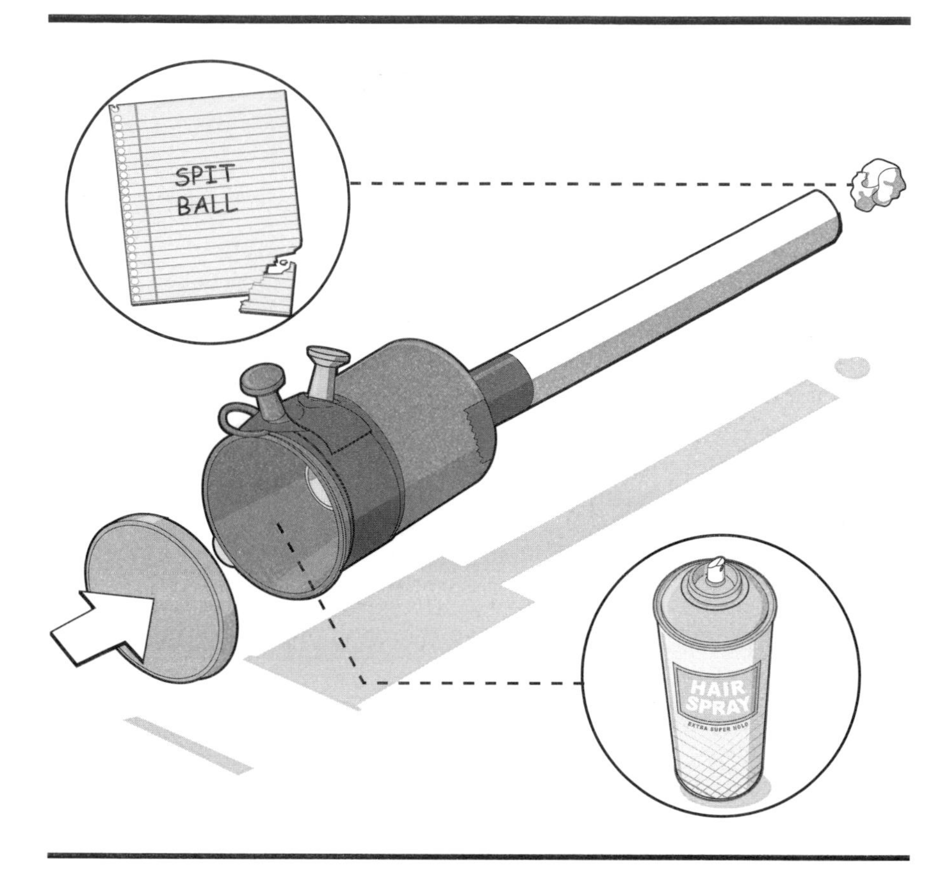


Now for the easy part, create a spitball in your mouth. Push the wet ball into the muzzle end of the MiniWeapon's barrel. Use the pen's ink cartridge to help push the spitball to the rear of the barrel but not into the combustion chamber.

Now add the aerosol hairspray (which contains alcohol, propane, or butane) to the combustion chamber. (Only a small burst of hairspray is needed for this microlauncher.) Then quickly cap the chamber and aim it carefully. Trigger the ignition switch and watch the spitball fly.

Just like with every spitball, your results can vary using the Mini Spud-and-Spit. Variables include container size, barrel length, spitball size, hairspray amount, and the specific mixture of aerosol used in the hairspray.

# PING-PONG ZOOKA

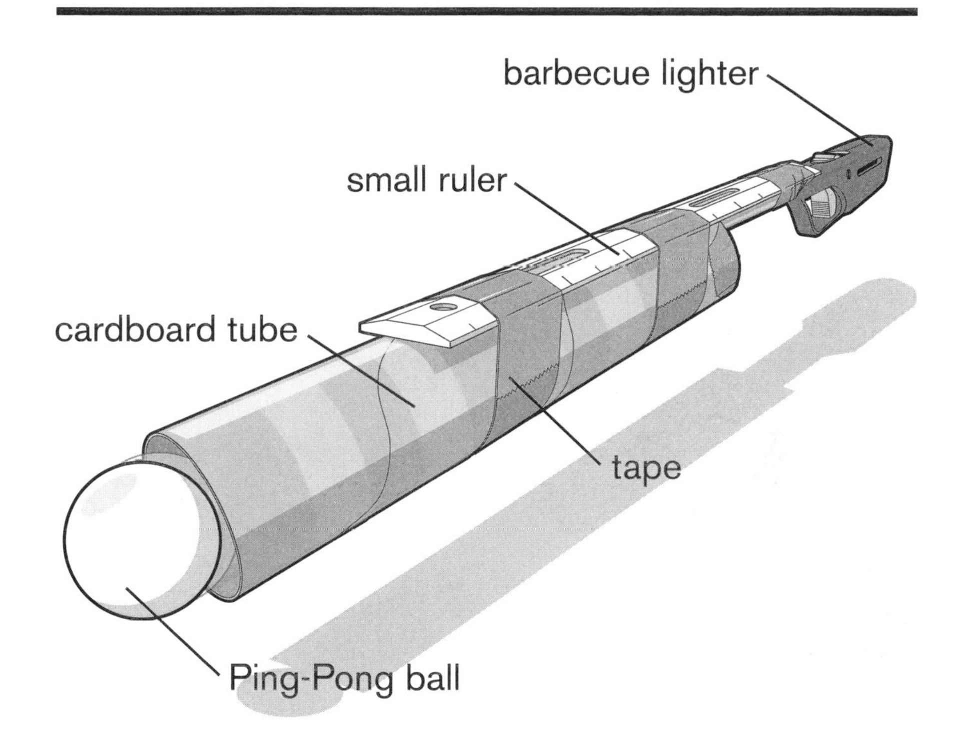


**Range: 20 feet**

The Ping-Pong Zooka is one of the only launchers you're likely to see that sends flames flying out of its cardboard muzzle. The zooka is a low-range combustion shooter with nonthreatening Ping-Pong ammunition. With its quick build time, you'll soon be launching balls of fury!

## Supplies

1 cardboard tube
Duct tape
1 barbecue lighter
1 small ruler
Hairspray

## Tools

Safety glasses
Hobby knife

## Ammo

1+ Ping-Pong balls

# Step 1

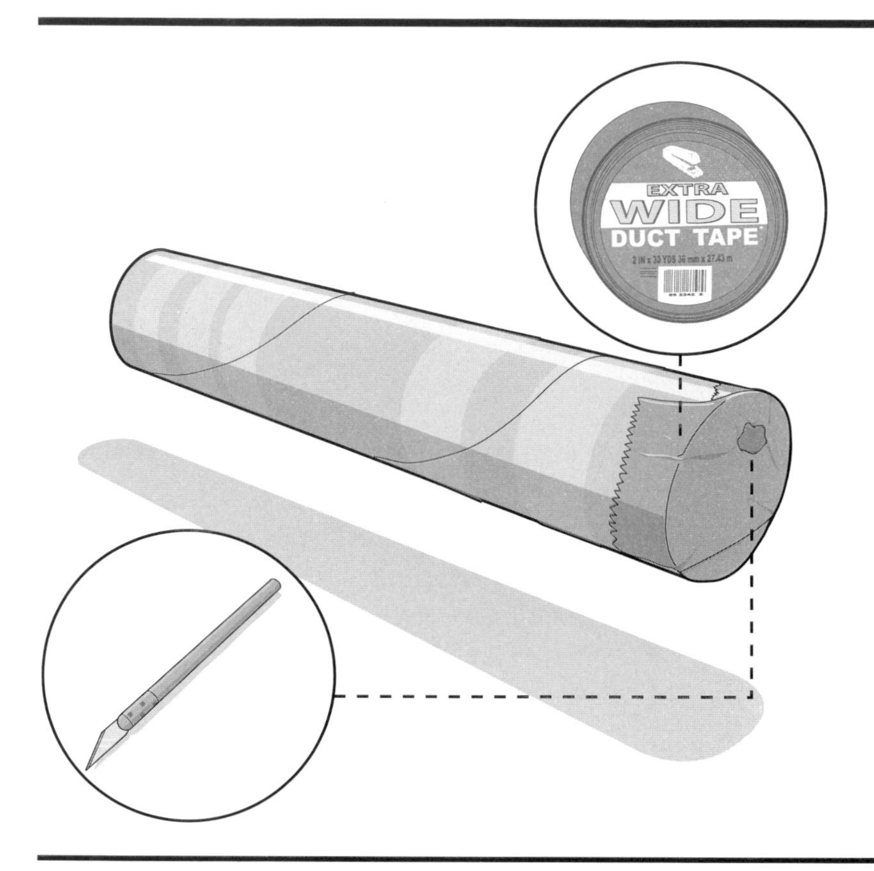


Paper towel cardboard tubes vary in size, so before you begin construction, check the diameter of your tube against the circumference of the Ping-Pong ball. It should be a snug fit.

Using duct tape, cover one end of the paper towel tube. Now, with a hobby knife, create a small incision in the duct tape cover. The diameter of this hole should be the same size as the nozzle on the barbecue lighter you'll be using.

## Step 2

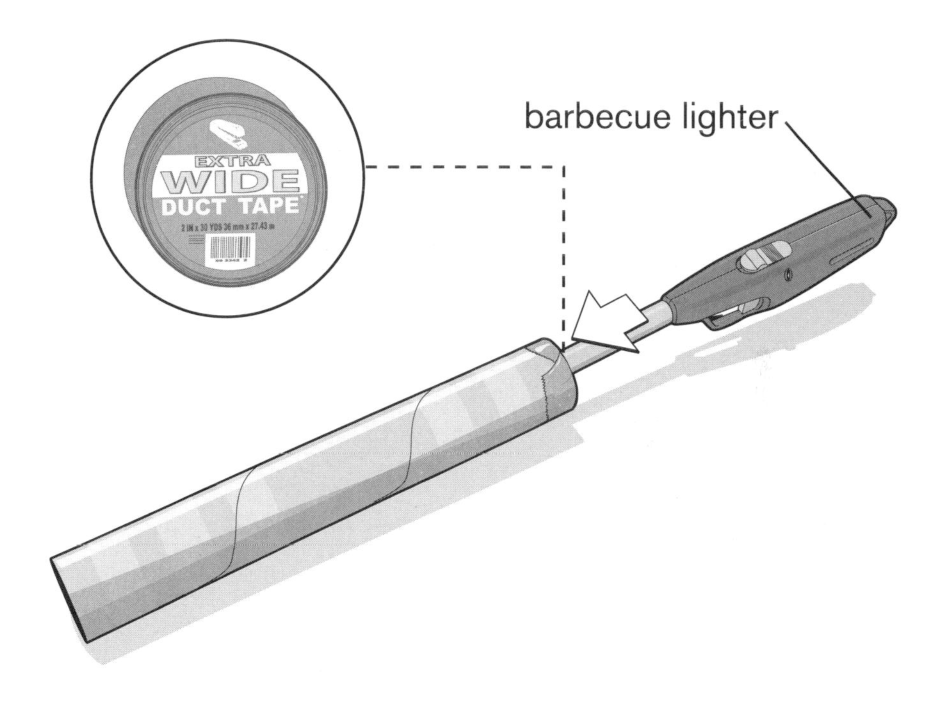


Install the zooka's ignition source by inserting the lighter nozzle through the duct tape incision at the end of the cardboard tube. The nozzle should extend 1 to 2 inches into the tube.

Use duct tape to seal any opening remaining around the tape incision. This seal will prevent the loss of combustion, so use as much tape as you need to ensure a tight seal.

# Step 3

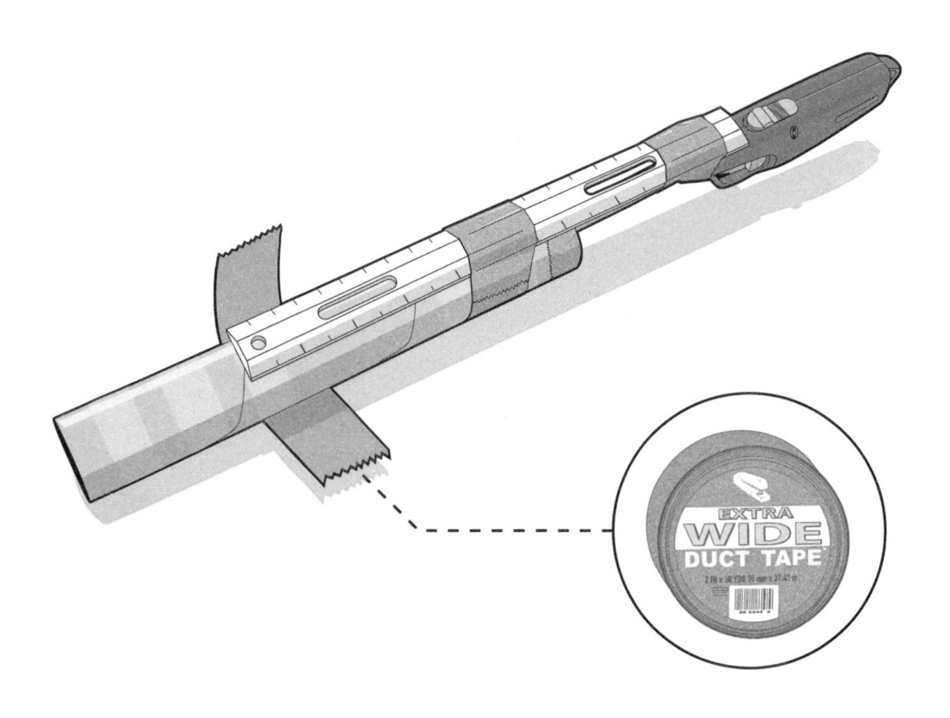


To add some much needed support to your Ping-Pong Zooka, duct tape a plastic or wooden ruler on top of the assembly. It should bridge the tube and lighter assemblies. This will allow you to keep one of your hands free while loading and firing.

## Step 4

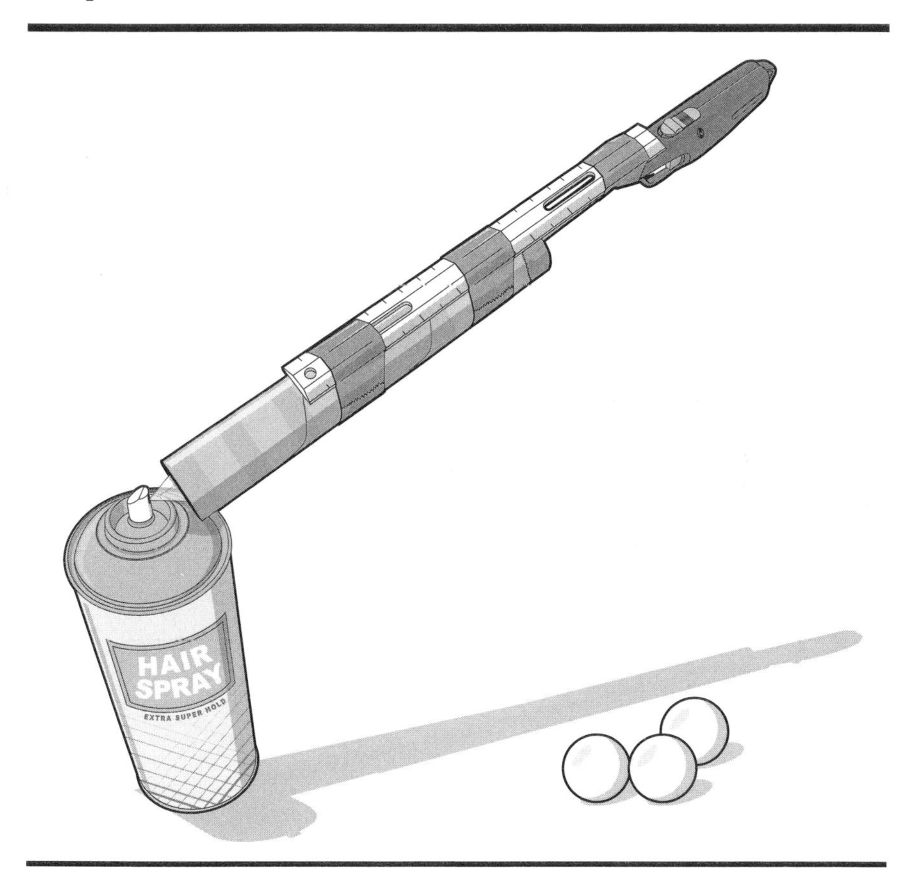


Hold your Ping-Pong Zooka so that the muzzle end of the barrel is positioned lower than the combustion chamber. Shoot a small amount of a slightly flammable aerosol hairspray or body spray (should contain alcohol, propane, or butane) into the cardboard tube. The flammable vapor will make it to the rear of the combustion chamber, right were you need it.

Immediately after spraying the aerosol into the cardboard barrel, insert the Ping-Pong ball. Safely aim your launcher and press the trigger. It is a good idea to add a small amount of hairspray the first time and then gradually increase it until you find a safe and appropriate amount to launch your Ping-Pong ball.

# 6

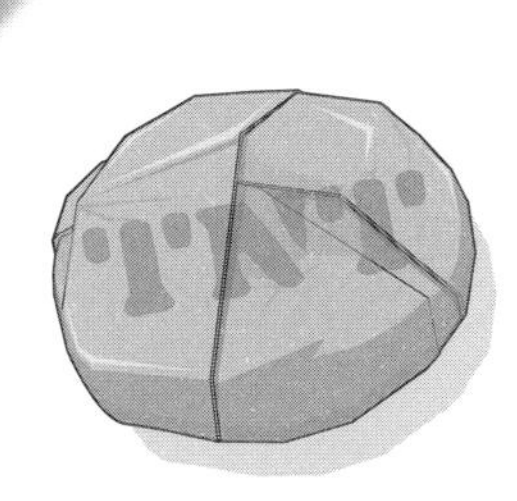


# MINIBOMBS AND CLAYMORE MINE

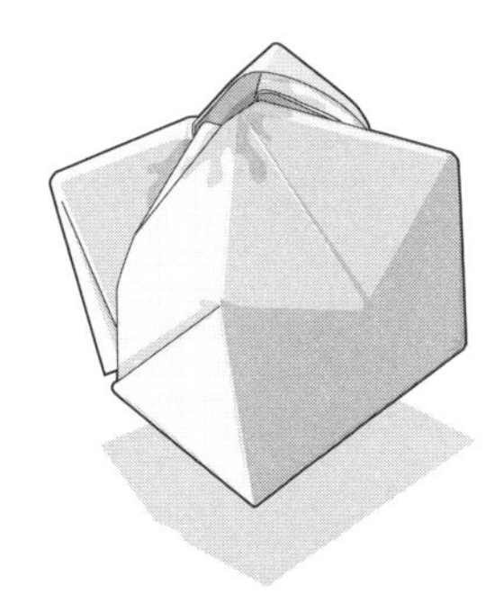

# MATCHBOX BOMB

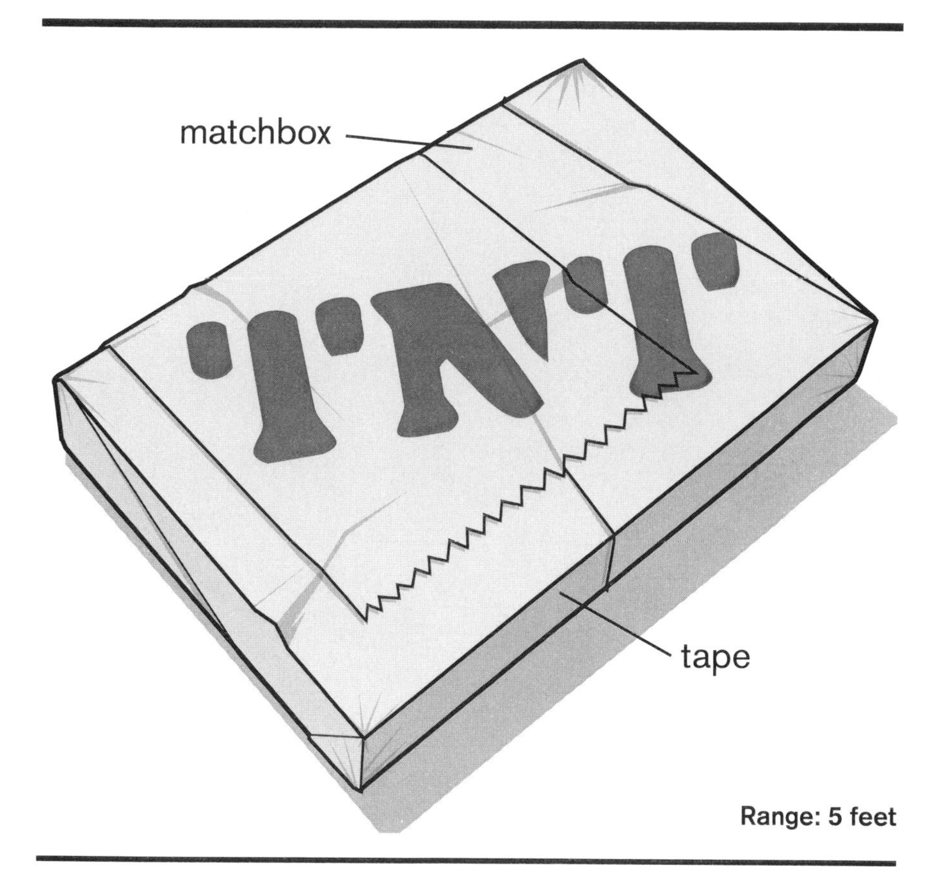


**Range: 5 feet**

This mini matchbox marvel will certainly have your neighbors jumping for cover. Designed to be thrown, the matchbox bomb is a perfect ear-piercing weapon. The universal design makes it ideal for any soldier to experiment with, using different-sized boxes to meet his or her mission's specific goals.

### Supplies

1 box of matches
Masking or duct tape

### Ammo

Matches

### Tools

Safety glasses
Earplugs
Scissors

## Step 1

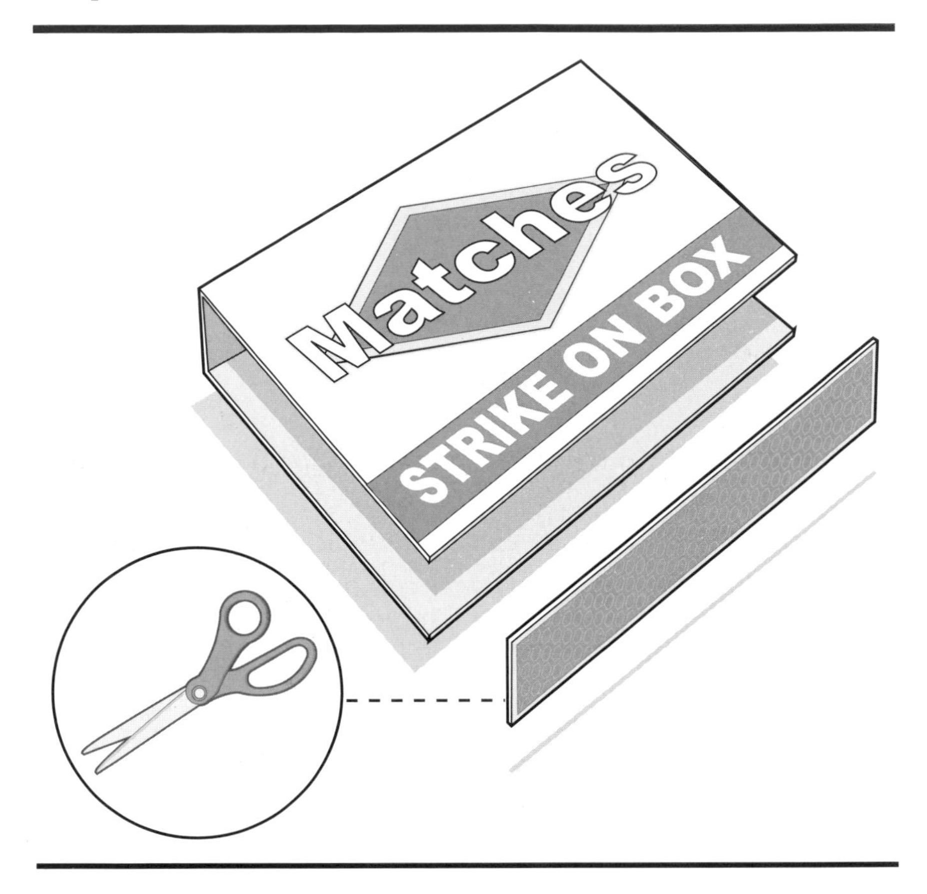


For this minibomb, you will need a box (any size) of wooden stick matches with a striking surface.

First, remove the cover from the matchbox. Use scissors or a hobby knife to slice off the striking surface on the side of the box. Save both the striking surface and the remaining box for the next steps.

## Step 2

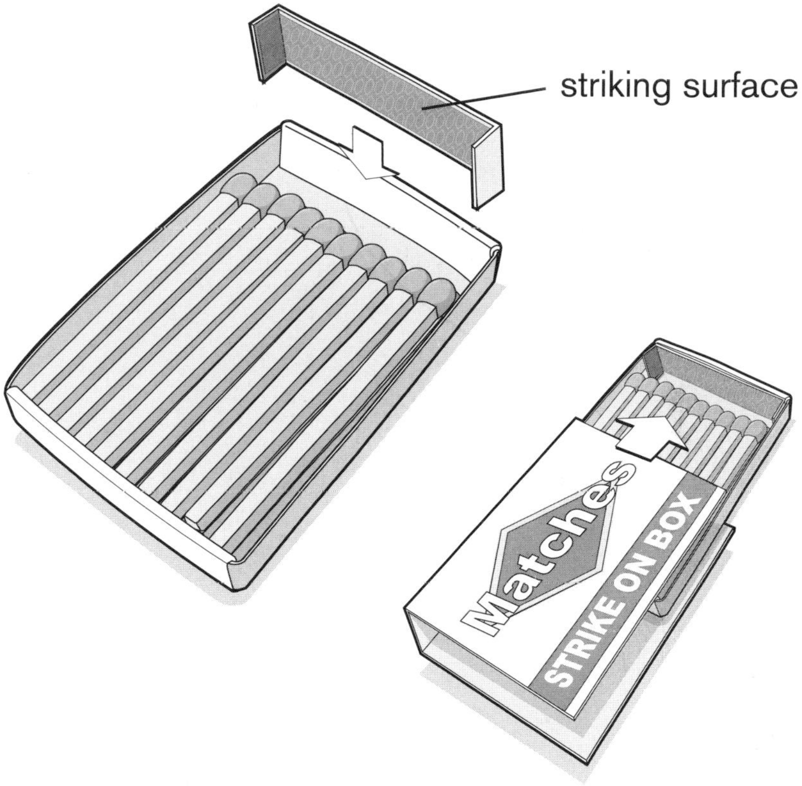

Once the striking surface has been removed, slide it in front of the match heads inside the box. Slide the remaining box cover over the box of matches to conceal the contents and keep the matches from falling out.

# Step 3

Tape the matchbox securely, using several layers of tape if necessary, to confine its explosive contents. Once wrapped, throw the box with all your might against a hard outdoor surface. The phosphorus match heads will ignite when they rub the striking surface. The gases released from the combusting matches will need to escape, causing a small-scale explosion.

As always, remember: ***safety first!*** Keep the Matchbook Bomb away from flammable materials and never use it indoors. The matchbox can be unstable, so handle it with care before detonation. This Matchbox Bomb is also capable of producing loud noises, so earplugs and eye protection are recommended. Use at your own risk.

# PENNY BOMB

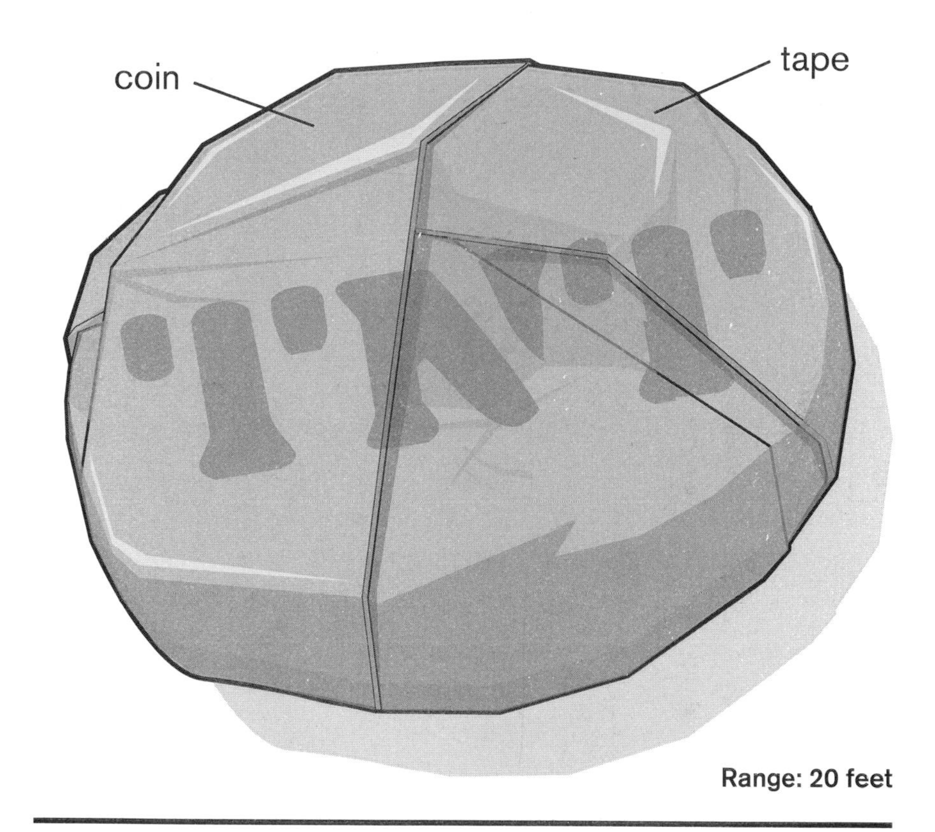


Range: 20 feet

Penny Bombs get their explosive power from low-grade Armstrong's mixture paper, better known as paper caps. Caps for toy guns are available in large rolls of 500, which will have you blasting every penny you have. Once you learn to build these sound grenades, you'll be covering your ears in no time.

## Supplies

1 roll of paper caps
Transparent or masking tape

## Ammo

1 penny

## Tools

Safety glasses
Earplugs

# Step 1

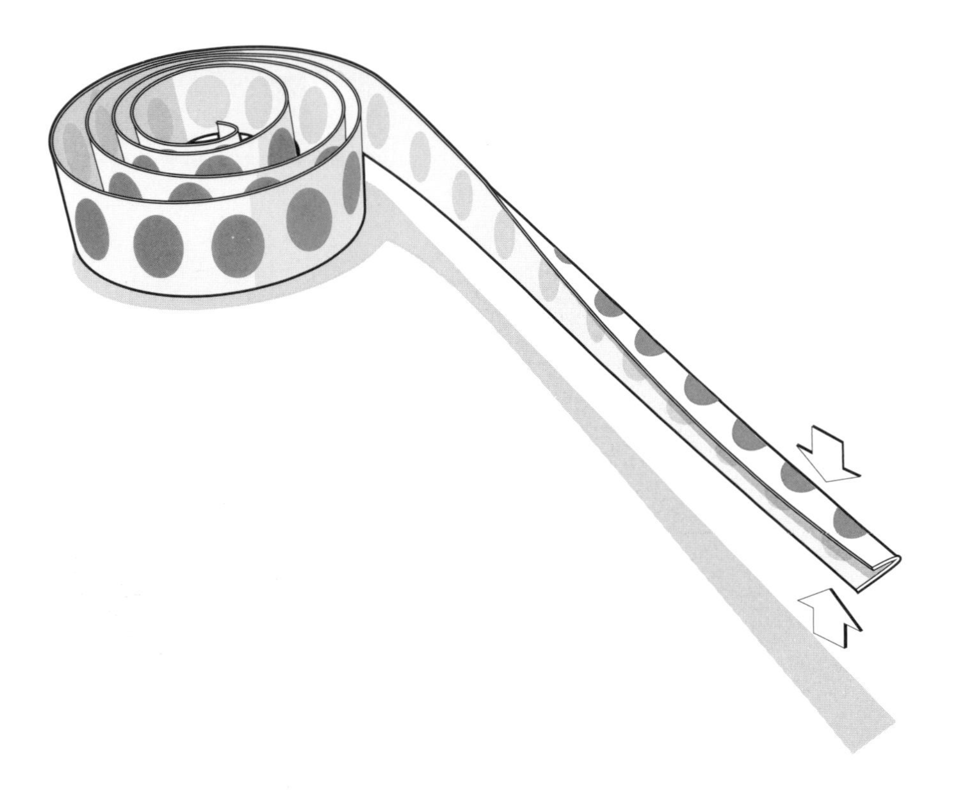

Unroll your caps on a smooth surface. Then fold the paper down the middle to reduce the width by half. The number of caps you use is your decision; however, 100 caps is more than adequate for a single Penny Bomb.

## Step 2

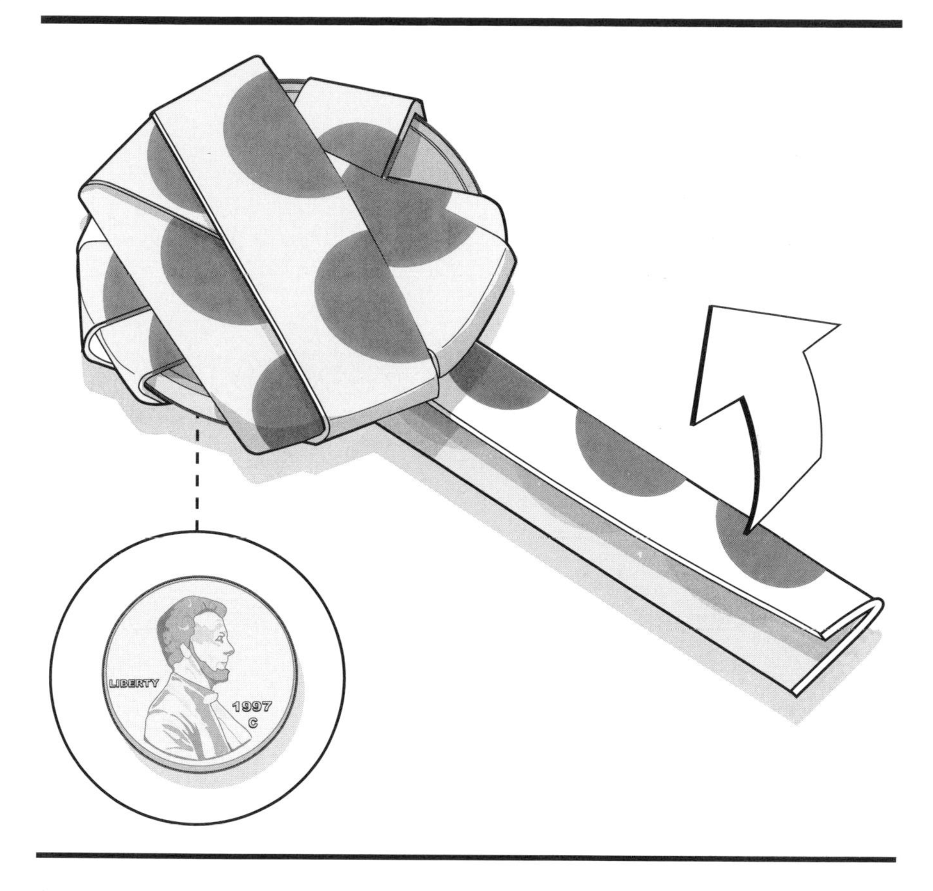


Take the folded paper caps and tightly wrap them around a penny several times, overlapping each previous layer until you've dramatically increased the size of the coin. The coin will add much needed weight to the Penny Bomb when you throw it.

## Step 3

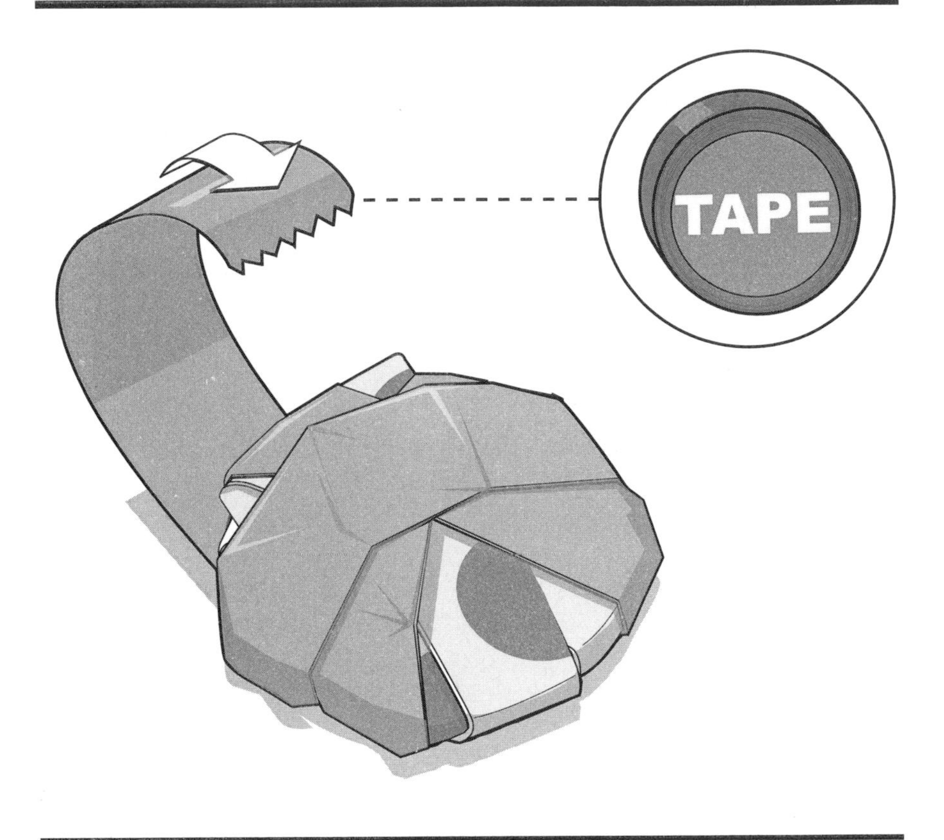


Once the folded caps are wrapped around the coin, tape them in place. In this case, the less tape you use the better.

Take your minibomb outside and throw it against the pavement. The Penny Bomb is capable of producing loud noises, so ear and eye protection are recommended. Always keep safety in mind, and never throw a Penny Bomb at anyone.

# WATER BOMB

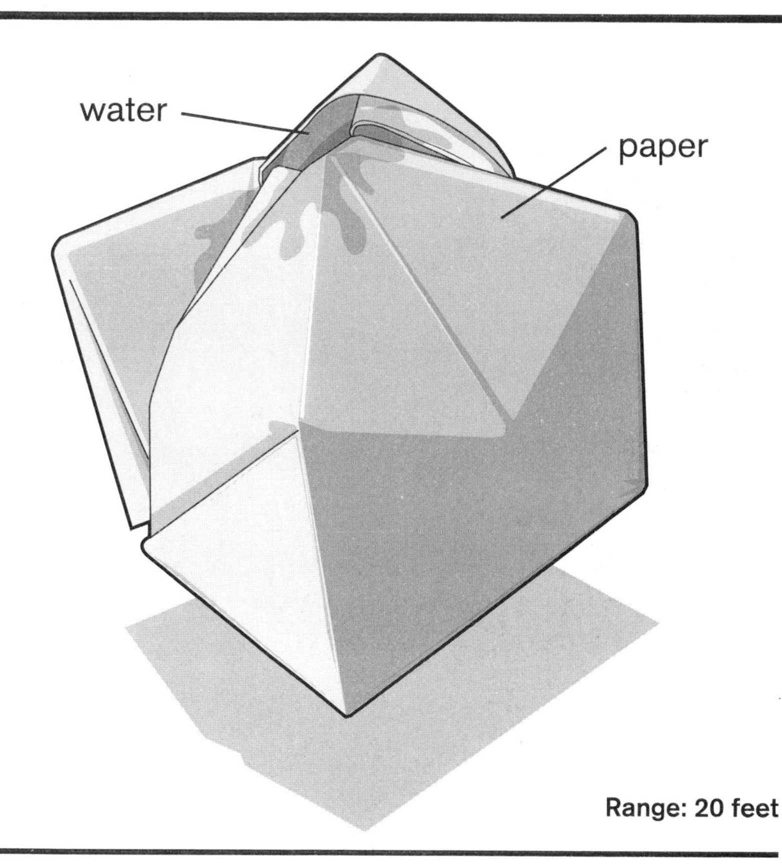


Range: 20 feet

The Water Bomb is constructed using only one sheet of paper. You create it by using your origami skills to form a small container capable of holding water. Once filled, time is of the essence. Throw your water grenade at the unsuspecting target, and on impact it will break apart, drenching your victim.

### Supplies

1 sheet of paper

### Ammo

Water

### Tools

Scissors

## Step 1

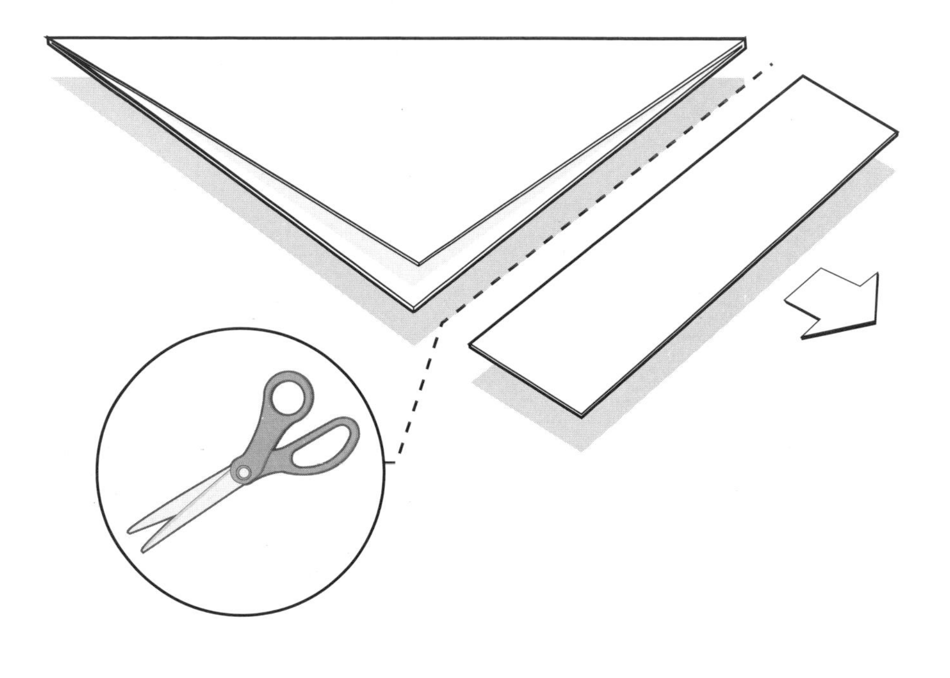

Take a standard-sized piece of paper and fold it diagonally as shown. Then use scissors to trim off the extra rectangle. Discard the trimmed piece and unfold the triangle to find a perfect square of paper.

## Step 2

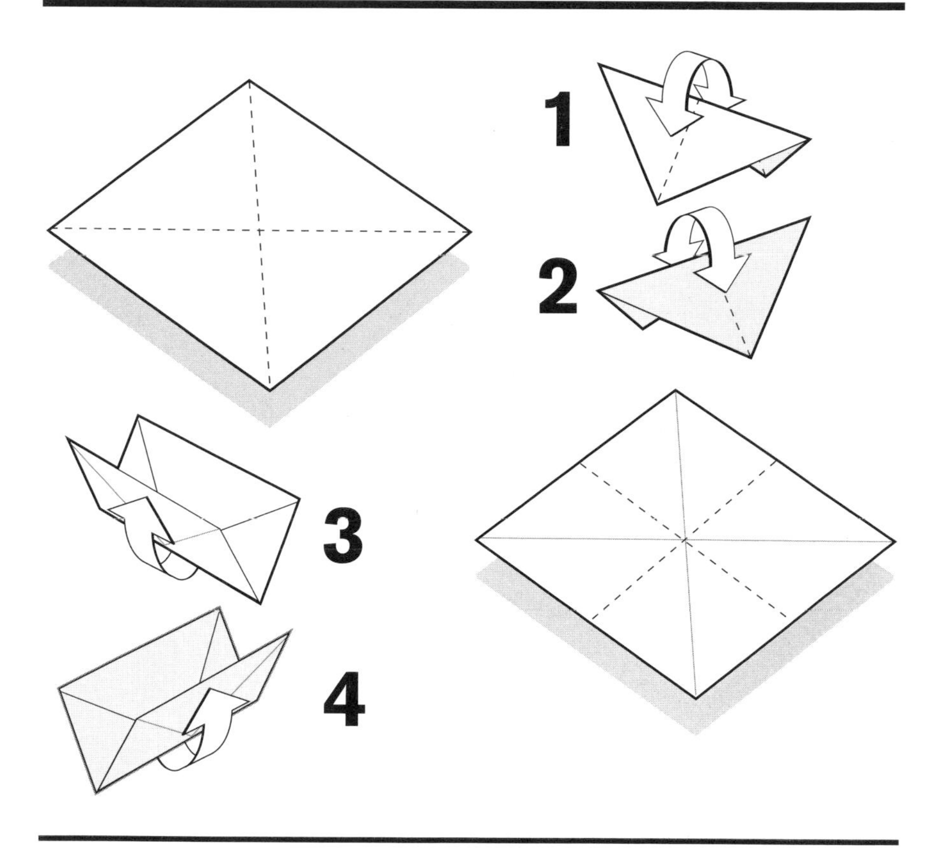


Now for some creative folding. Fold the paper along the other diagonal and then in half both ways until you have what appears to be a star pattern. Use the illustration above for reference. This may take a few tries due to the confusing nature of the creases.

## Step 3

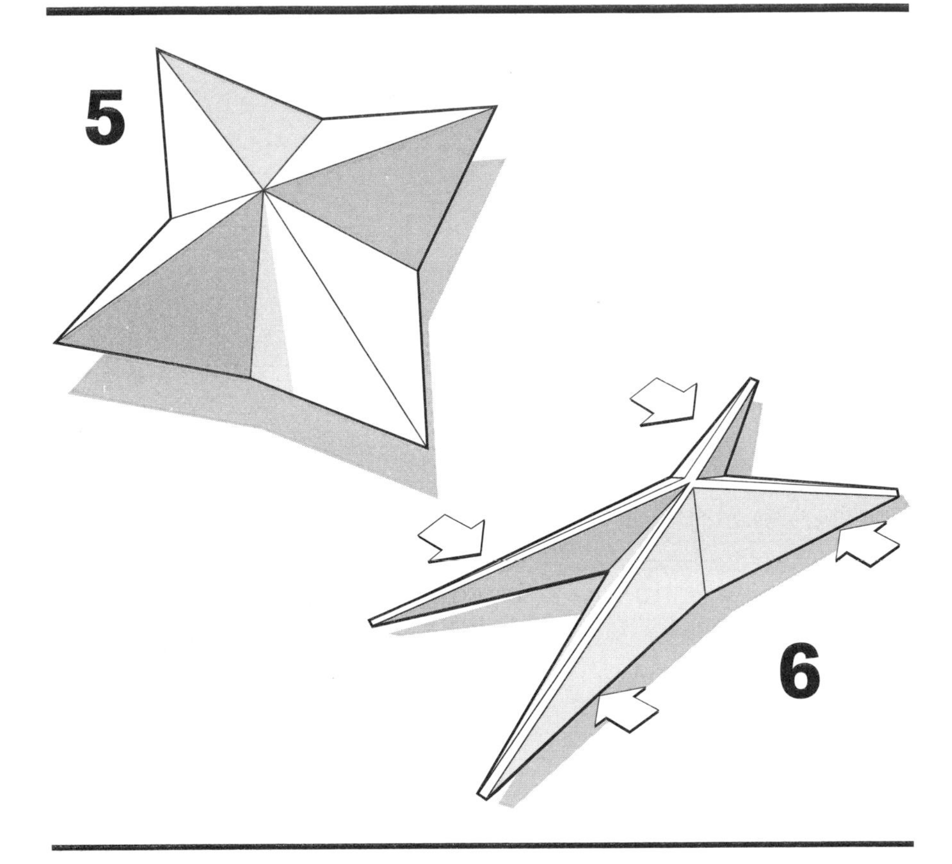


Pull the center point forward while pushing in at the center of the four edges to form a star. Once this is completed, run your fingers along the fold lines to crease the edges. This will help the paper keep its form during construction.

Finally, push in the two adjacent sides to form a quadruple-walled triangle.

## Step 4

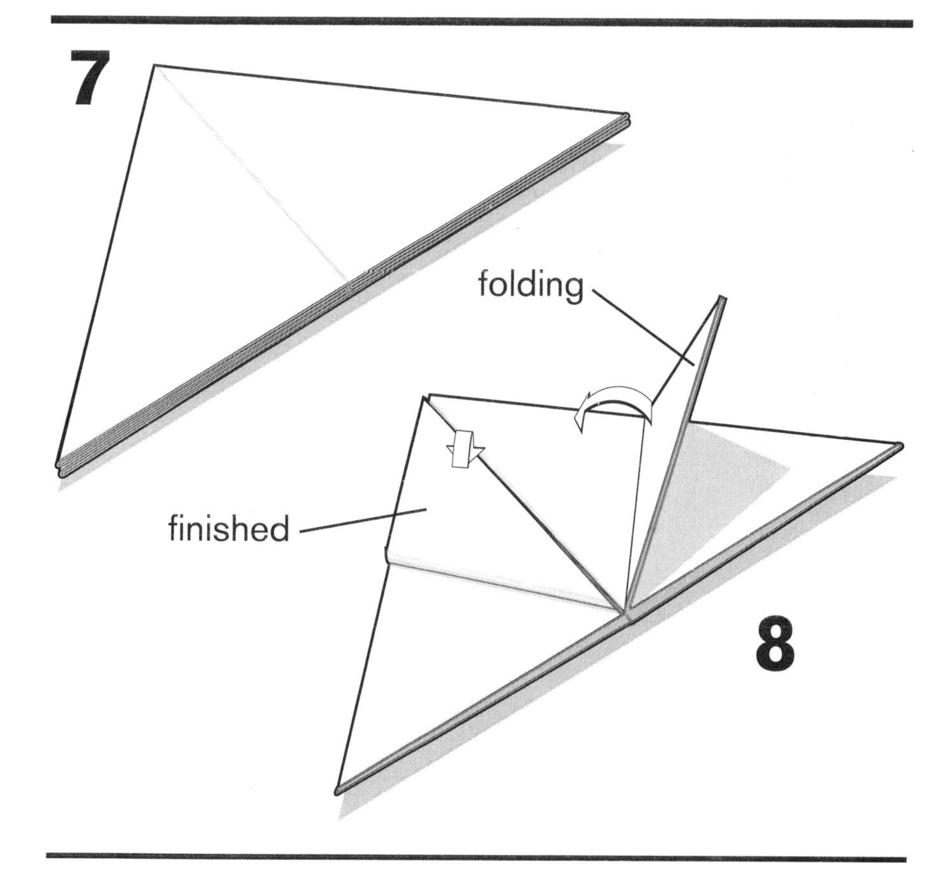


Next, take the thick paper triangle and lay it flat with the bottom edge facing you.

Fold the bottom right corner toward the center to form a smaller triangle. The corner should reach up to the top point. Repeat this step with the bottom left corner. When finished you will have created a diamond shape on this side. Crease the edges with your finger.

## Step 5

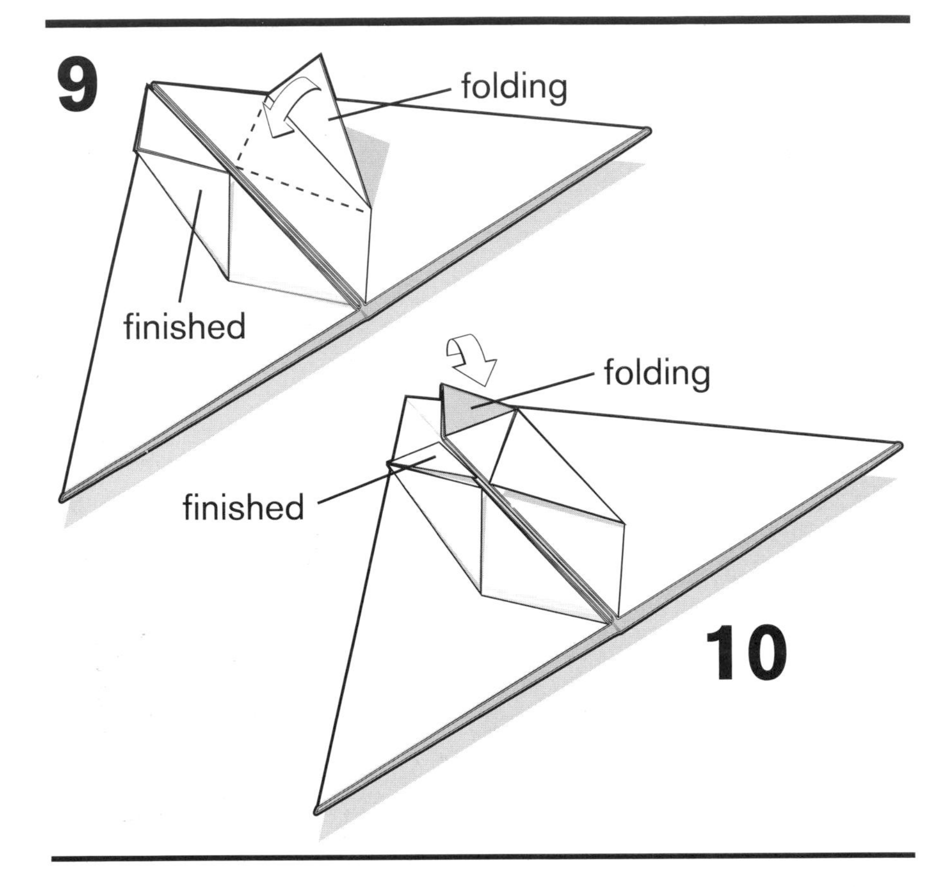


Take the left and right points of your diamond and fold them in, to the centerline, creating two smaller triangles. See illustration for reference.

Next, fold down the top of these smaller triangles so they are flush with the other two triangles, as shown.

## Step 6

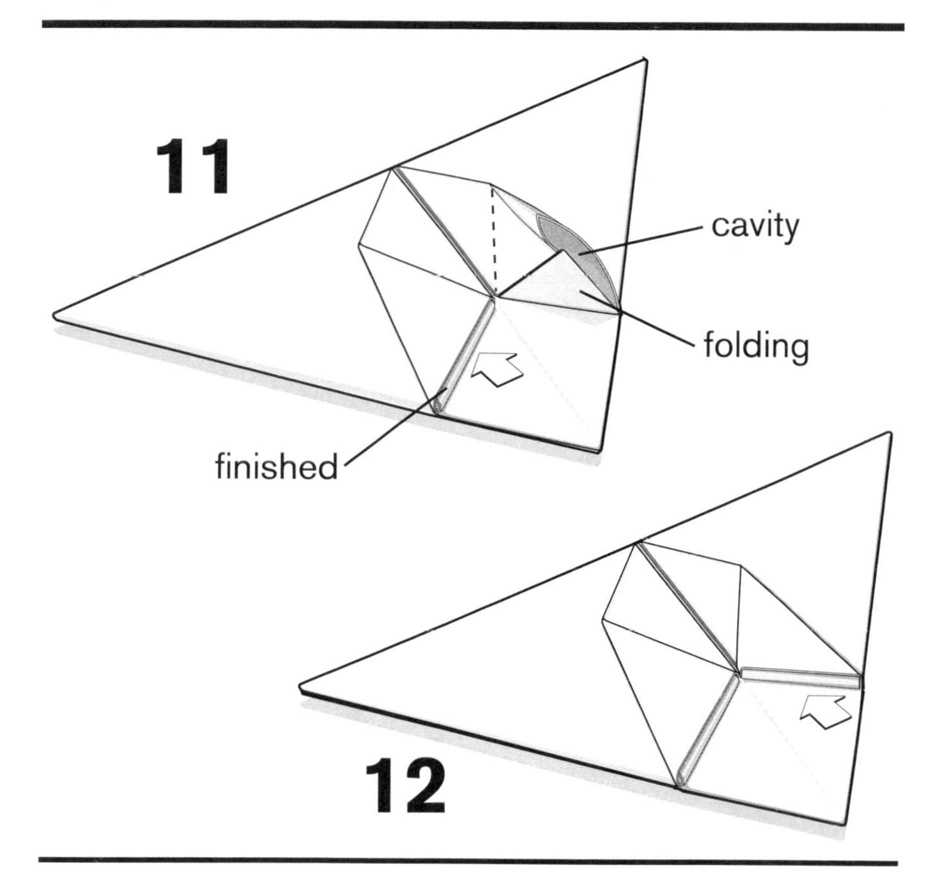


Rotate the triangle so that the point faces you.

Fold one of the smallest of the triangles into the cavity right below it (which was created when you folded in the side of the diamond earlier). It will take some work to get it nestled in place. Repeat this step for the adjacent side.

This completes the folding on this side. Now you will have to flip over the paper triangle and repeat steps 4, 5, and 6.

## Step 7

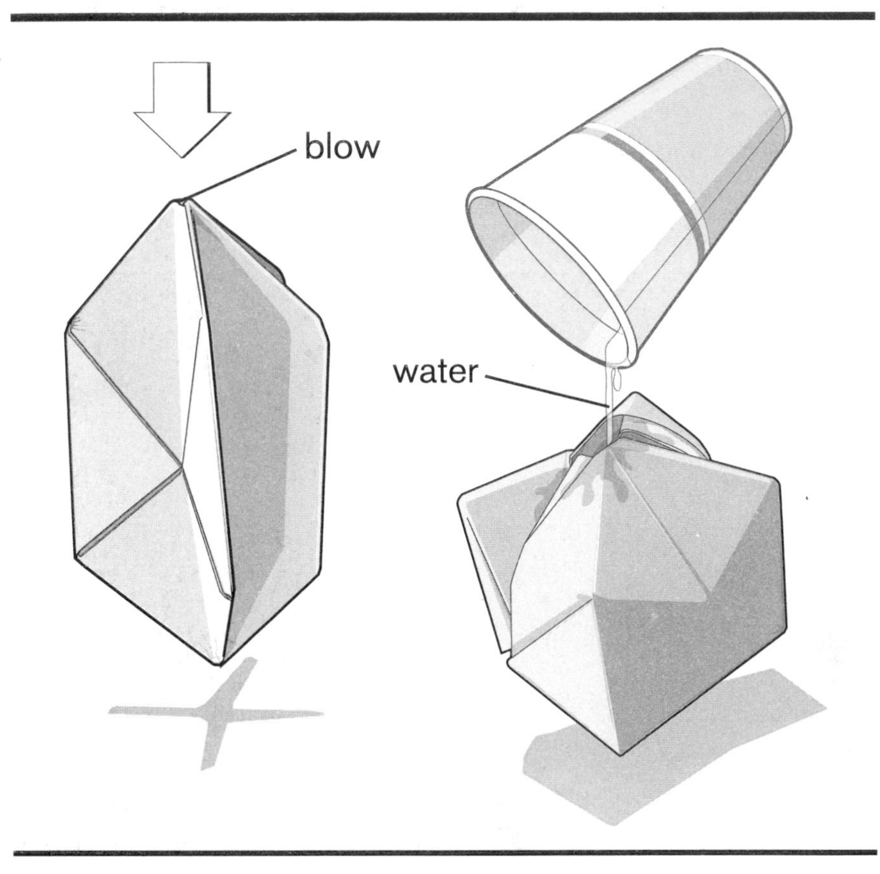


The Water Bomb is now finished—it's time to add air. Locate the end with the hole in it and blow as hard as you can to inflate the paper box. You may need to use your hands to gently help the paper balloon open and take shape.

When you know your target's arrival is near, fill the paper box with water and lie in wait. Once the target is in sight, launch your Water Bomb. It is important to keep in mind that the Water Bomb won't last long once it's filled with water, so you must act quickly. The slow grenadier gets wet.

# CLAYMORE MINE

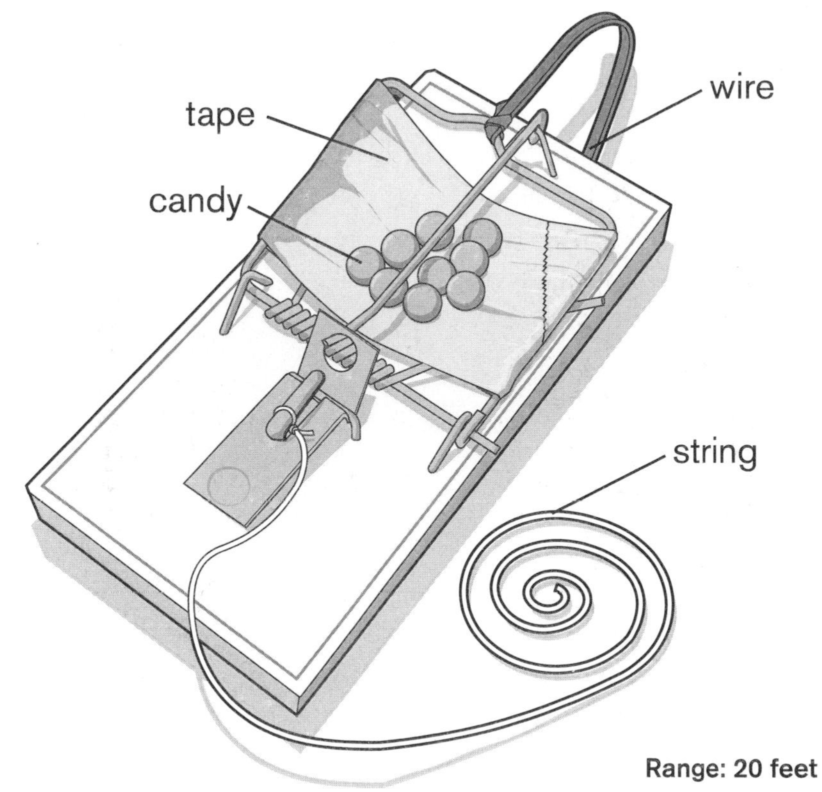


Range: 20 feet

The Claymore Mine is a self-setting miniature land mine that is capable of firing candy shrapnel—perfect for ambushing any cookie-stealing intruders. This basic design can be used for creating a perimeter matrix or scaring someone. It also makes an awesome tool during a game of cops and robbers.

### Supplies

Wire
1 mousetrap
Tape (any kind)
String

### Tools

Safety glasses
Wire cutter
Stapler

### Ammo

Candy

# Step 1

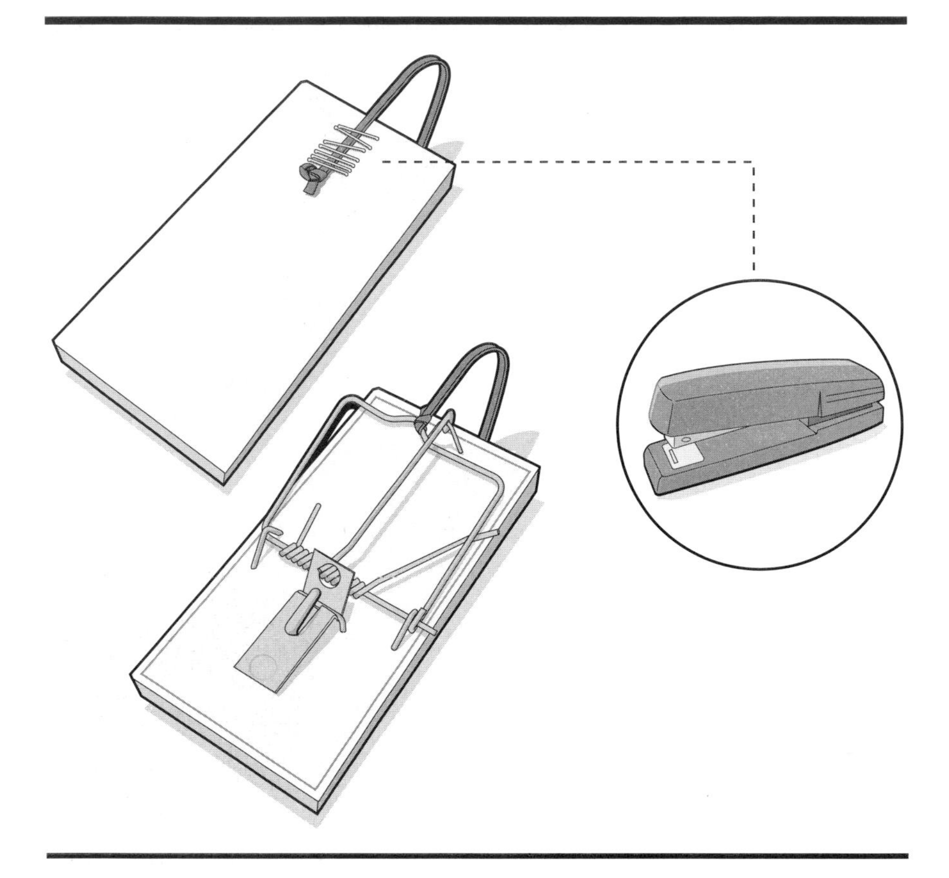

Cut approximately 6 inches of speaker wire to harness your spring mechanism. Tie one end of the speaker wire to the mousetrap bar. Then, using staples or a screw, fasten the other end of the speaker cord to the bottom of the mousetrap. Adjust the length of the cord so that, once set, the swinging mousetrap bar only rotates half or a quarter of the rotation to its normal, closed position. Where this bar stops will determine the direction your shrapnel flies.

## Step 2

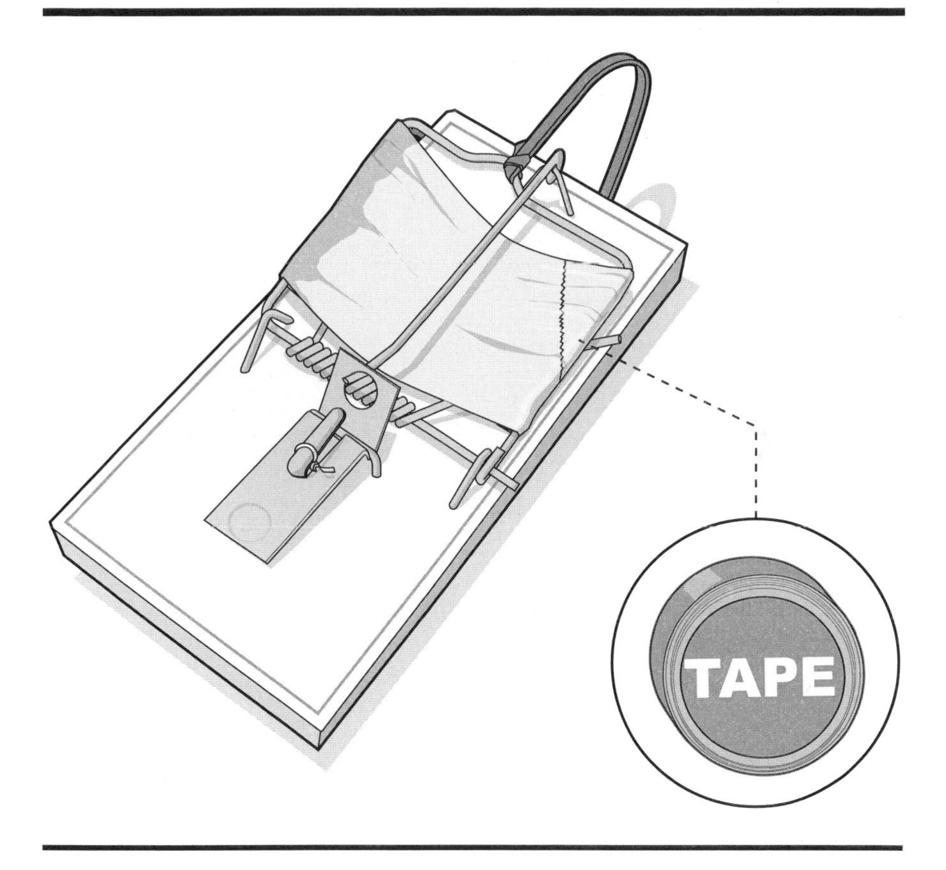


Holding the mousetrap face up, use tape to create a concave pouch on top of the mousetrap bar. This tape will act as the launching platform for your sugary shrapnel.

# Step 3

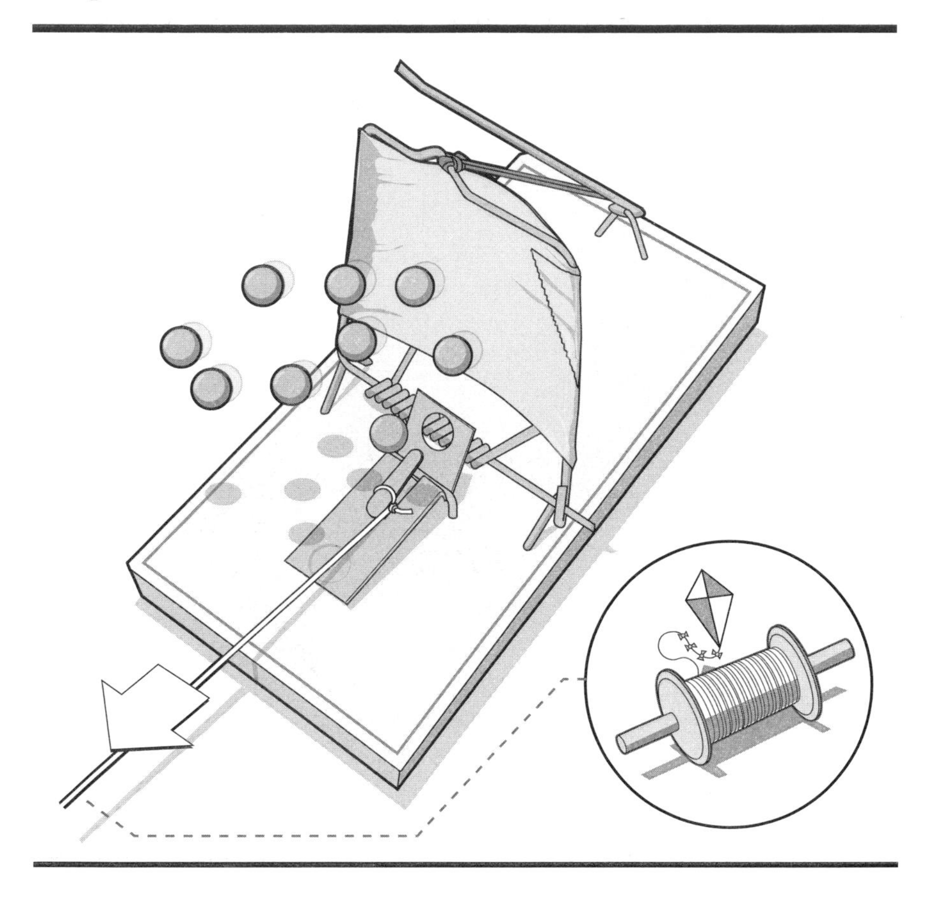

Strong string or high-test fishing line is perfect for the trip line. Tie one end of the line to the cheese trigger on the mousetrap. Now find a great location close to the ground and tape the mousetrap in place. With no slack, tie the other end of the line to something permanently stationary. Set your trap and carefully place your small candies onto the launch pad—then wait.

If using the Claymore Mine on the lawn or outside, add small spikes to your device by drilling large screws through the mousetrap's wooden frame and into the ground.

Always wear your safety glasses when playing with this Mini-Weapon. The shrapnel will fly in every direction, so use at your own risk.

## Alternate Construction

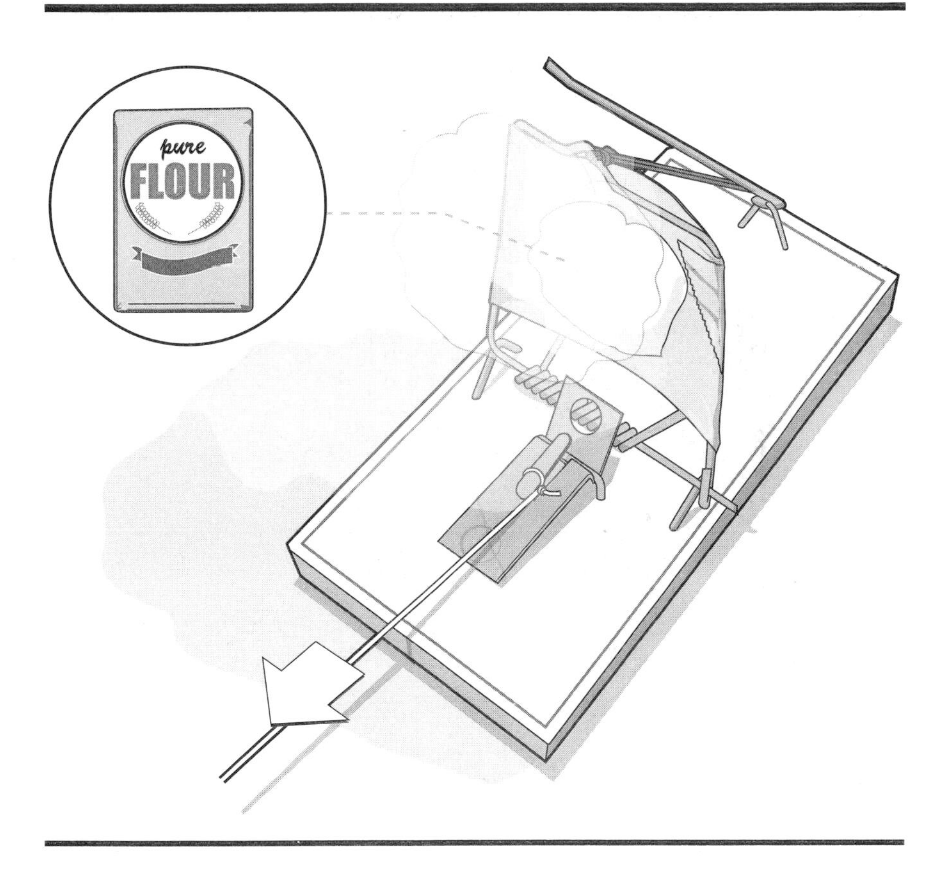


Looking for a realistic smoke explosion? Add baking flour to the shrapnel pouch. Once the line is hit and the shrapnel is released, the flour will go flying, making a dusty cloud.

This is a great addition to your indoor miniarsenal.

# 7

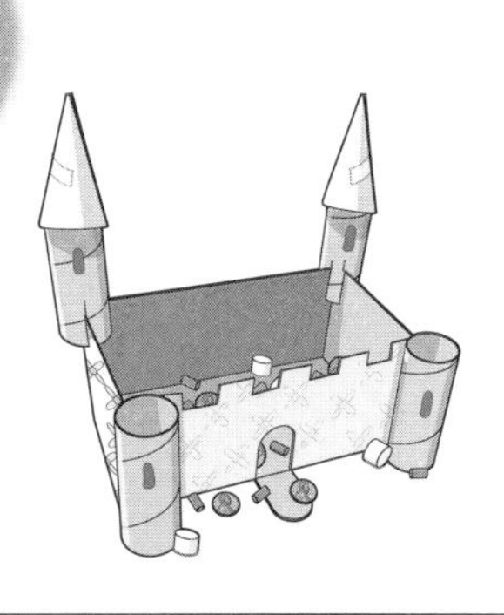

# CONCEALING BOOK AND TARGETS

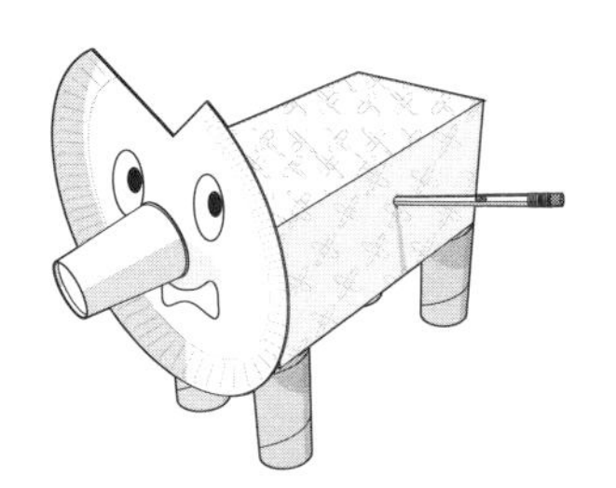

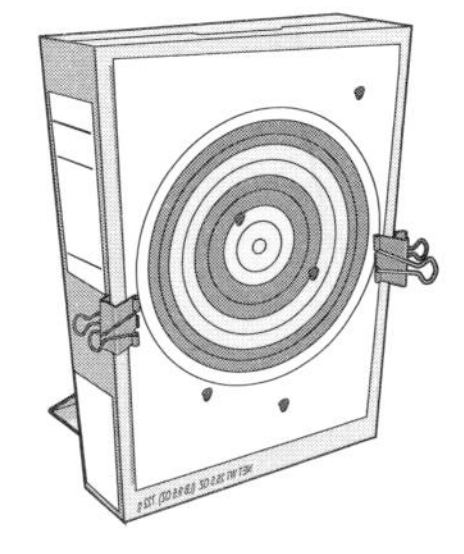

# CONCEALING BOOK

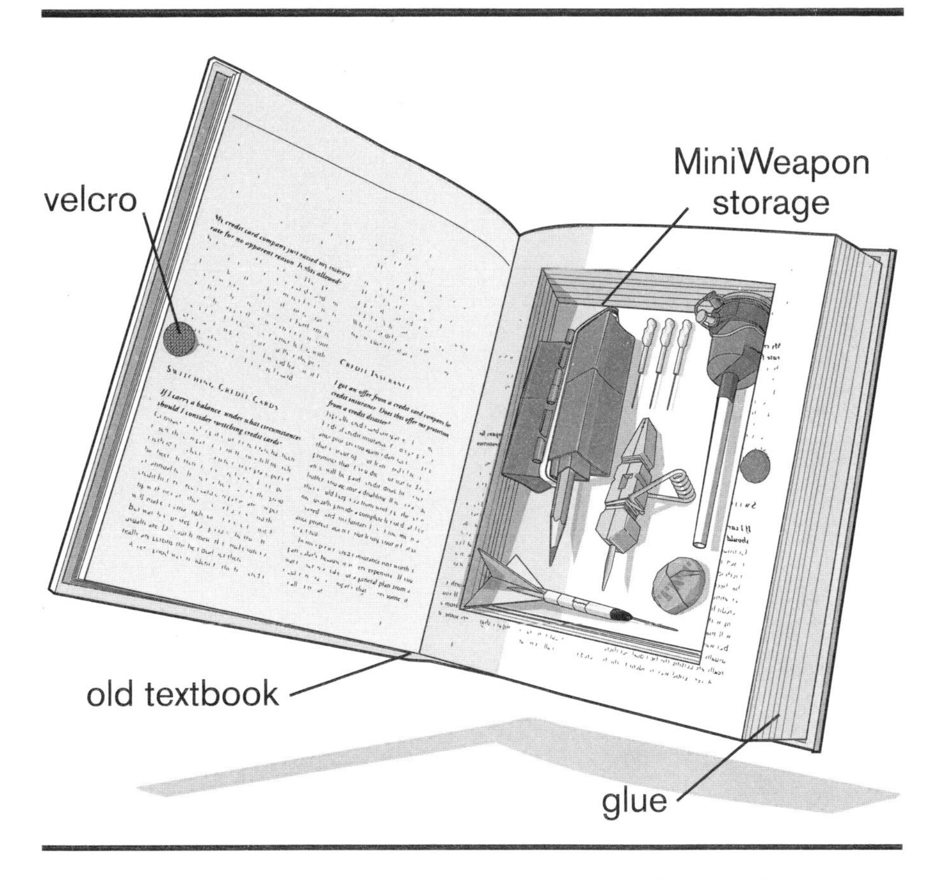


This hollowed-out book serves as a secret safe for stashing all your MiniWeapons of Mass Destruction. It's the perfect tool for unannounced room raids, and it can also be used to hide cash to finance your minirebellion.

## Supplies

1 old textbook
1 plastic freezer bag
White glue
1 paper or plastic cup
Water
Velcro

## Tools

Paintbrush
Ruler
Hobby knife
Pencil or pen

## Step 1

Place a plastic freezer bag as a marker somewhere in the last few pages of the textbook you plan to use. This will help control the running adhesive and protect your work surface.

Pour some white glue into a paper cup and mix in water—about two parts water to one part glue. Brush the mixture onto the sides of the textbook pages.

## Step 2

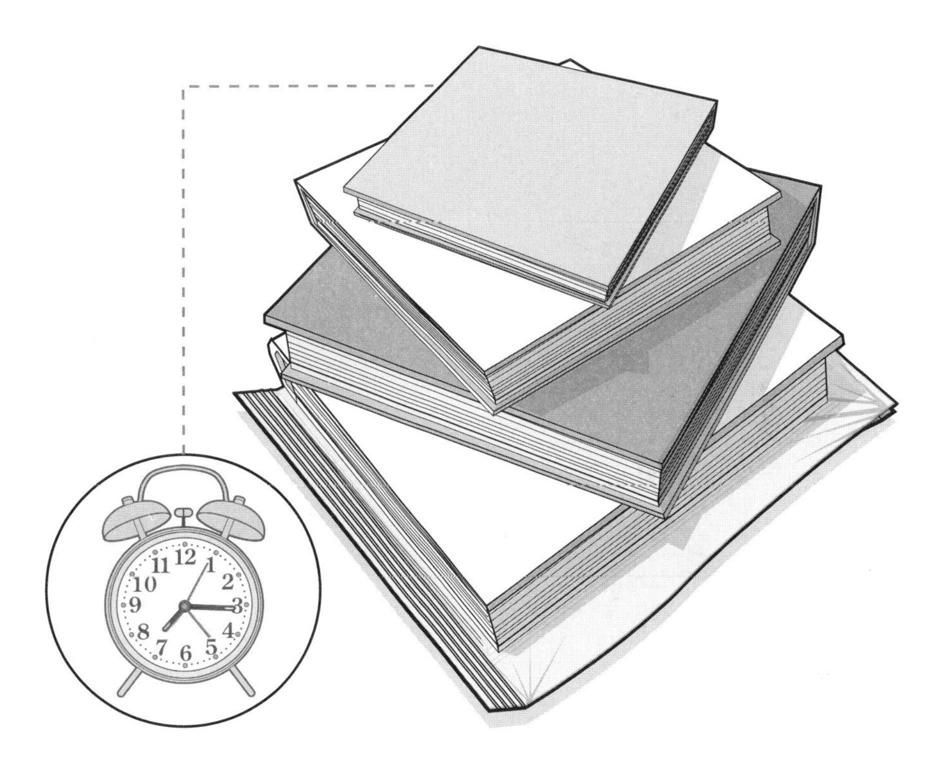

Once you've finished coating the sides thoroughly, let the glue dry. The damp pages will have a tendency to wrinkle while drying. To prevent this, stack something heavy (more textbooks will do nicely) on top of the drying book.

Once the adhesive is dry, go back and add a second coat. Allow it to dry, too.

## Step 3

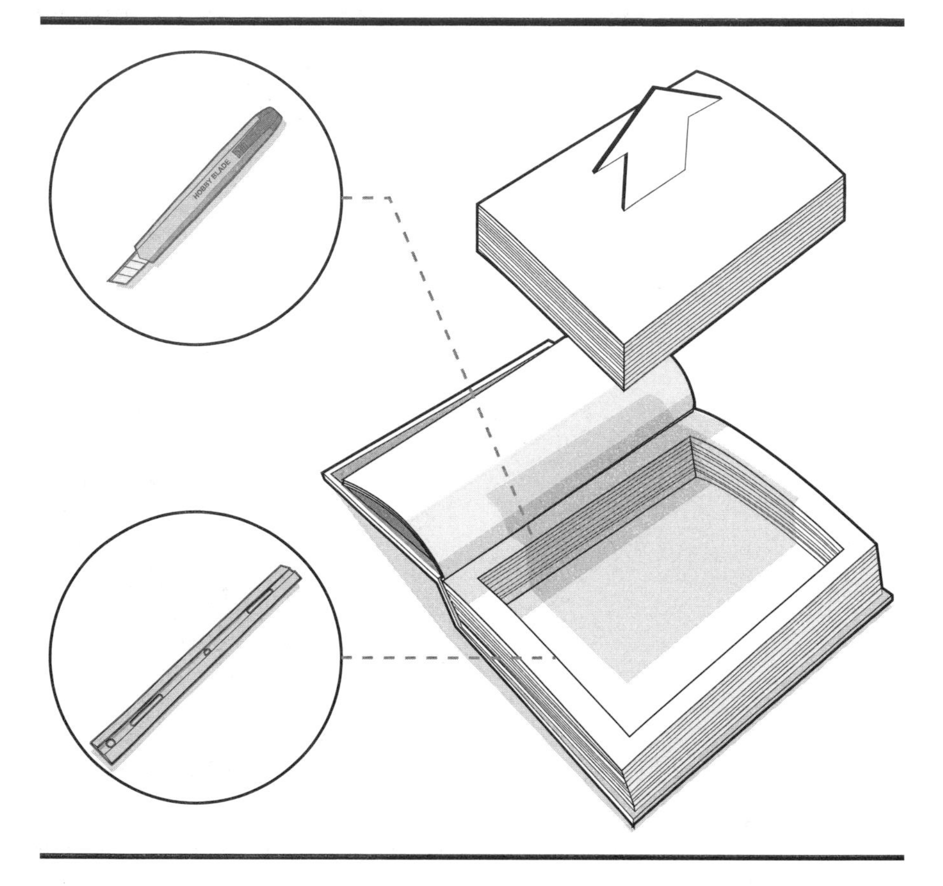


Now that your book has dried, it's time to cut out the center compartment. Use a ruler to mark out the compartment size you would like. Once you have your lines, use a sharp hobby knife to gradually cut out the layers of pages. Having the sides of the book glued will make this step easier.

Discard the page sections you cut out.

## Step 4

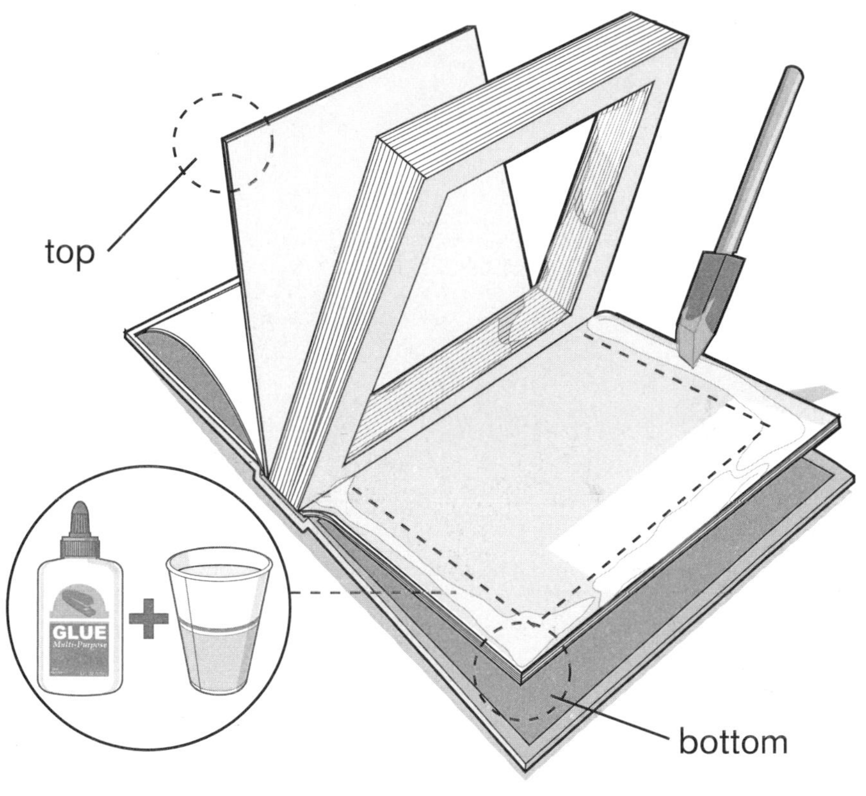


Now it's time to add the compartment bottom. You can either hot glue the compartment to the back cover of the book or use white glue to fuse together the last few pages of the book; a couple pages stuck together will form a solid bottom.

# Step 5

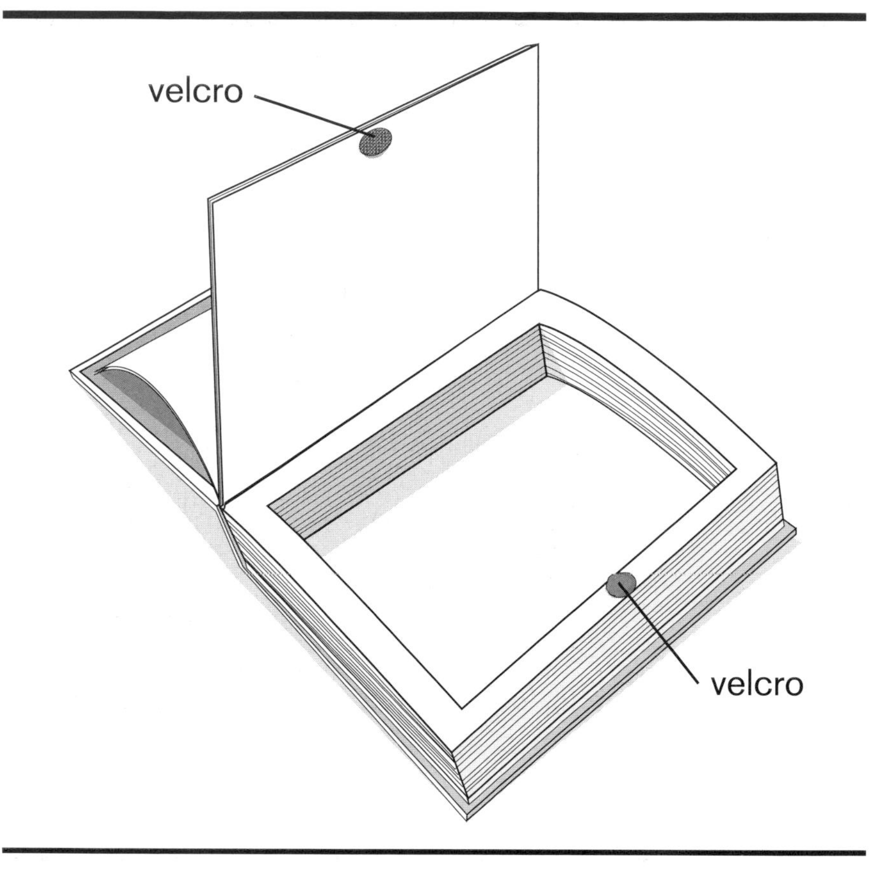


For extra security, add opposing pieces of Velcro with adhesive backing to your compartment cover as illustrated. This will help keep it closed when picked up. You certainly don't want to lose your precious cargo or expose your hiding spot.

# CATAPULT CASTLE

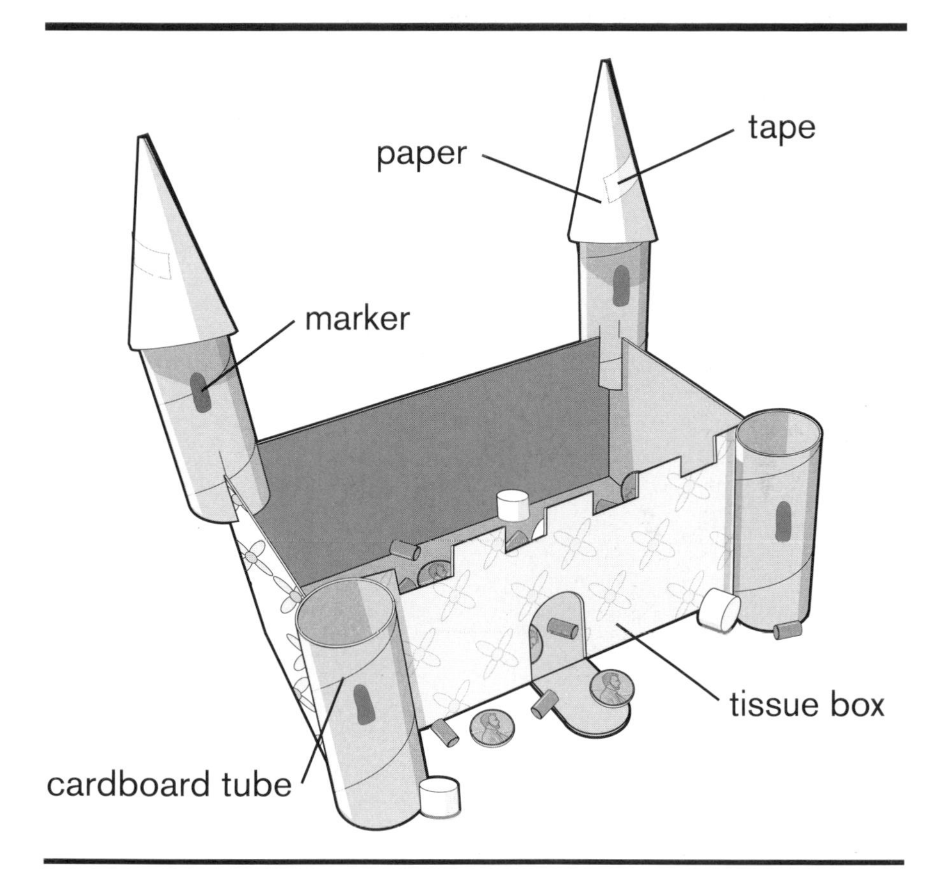


The tissue-box fortress is the perfect complement to your larger catapult arsenal. Its cardboard courtyard makes a great target for your mini marshmallow artillery. For game play, build two and out-aim your opponent.

## Supplies

1 tissue box
4 toilet paper tubes
Transparent tape
1 sheet of paper

## Tools

Scissors
Marker

## Step 1

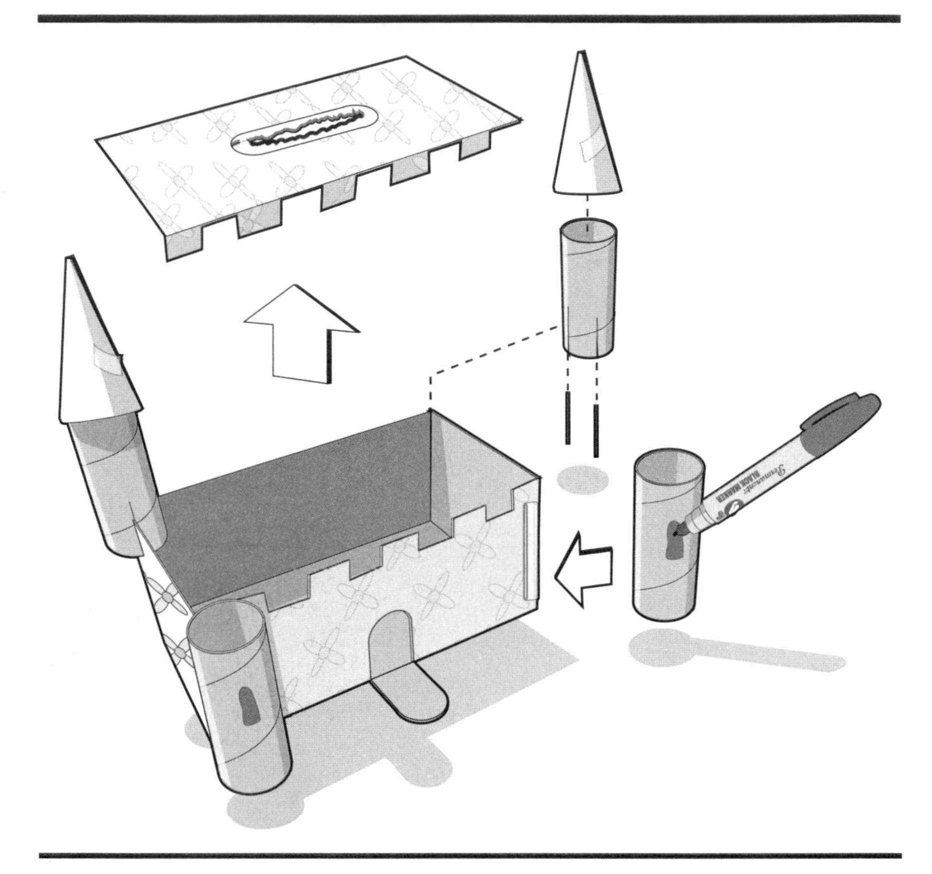

Remove the top of the tissue box. Add typical castle features such as battlements and a little drawbridge using your scissors. Then use toilet paper tubes for the four towers. The front towers can be attached with tape, and the back towers can be slid on after you've cut slits into the cardboard as shown. Roll two paper funnels, tape them in place on the rear towers, and cut them to size. With a marker, add window details to the towers and castle walls.

# BIRD SHOT

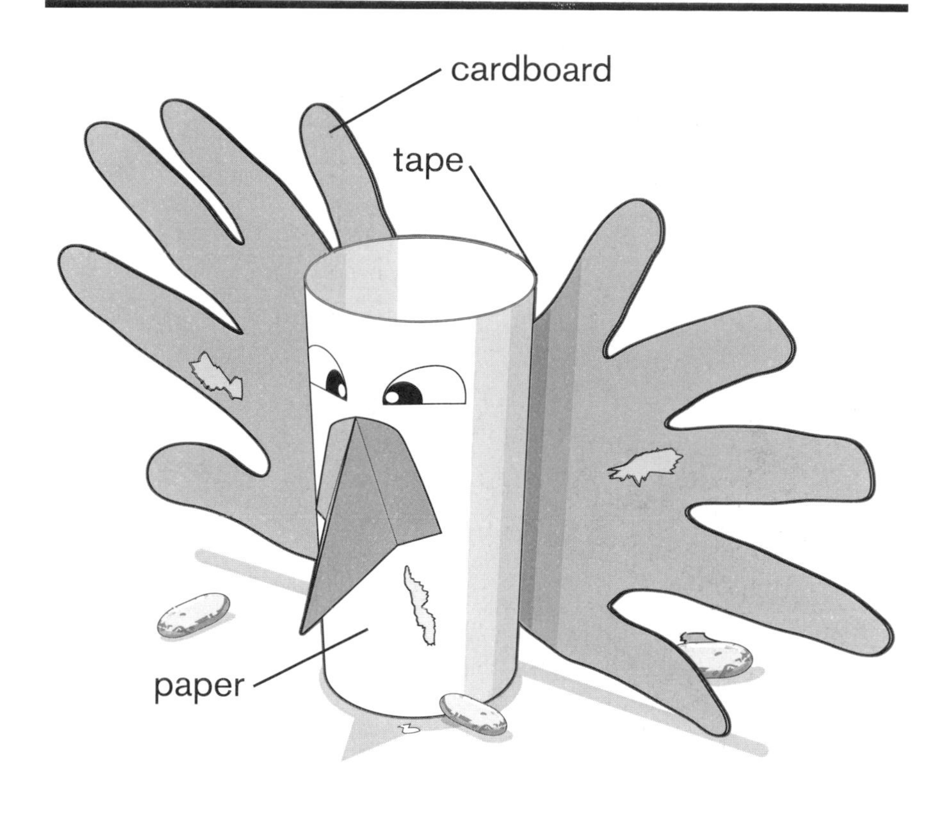


Quietly hiding in your bird blind, you patiently wait for the deadly man-eating cardboard bird to land within striking distance. You hear a noise behind you and before you know it, it's right on top of you! You spring for your Bean Shooter and unleash a hailstorm of birdshot to bring the beast down. Miraculously, you survive to tell this hair-raising story to your friends.

### Supplies

Cardboard
1 sheet of paper
1 aluminum can
Tape (any kind)

### Tools

Marker
Scissors

# Step 1

To create the Bird Shot's wings, trace your hands onto cereal-box cardboard and cut out the tracings with scissors. You will also need to cut out the flesh-eating bird's razor-sharp beak; this can be made with a simple folded triangle.

Next, roll a piece of paper around an aluminum can to form the bird's body. Once formed, tape the bird elements onto the paper cylinder. You can choose to leave the can inside the bird or remove it. Add scary eyes with a marker.

# ALIEN ATTACK

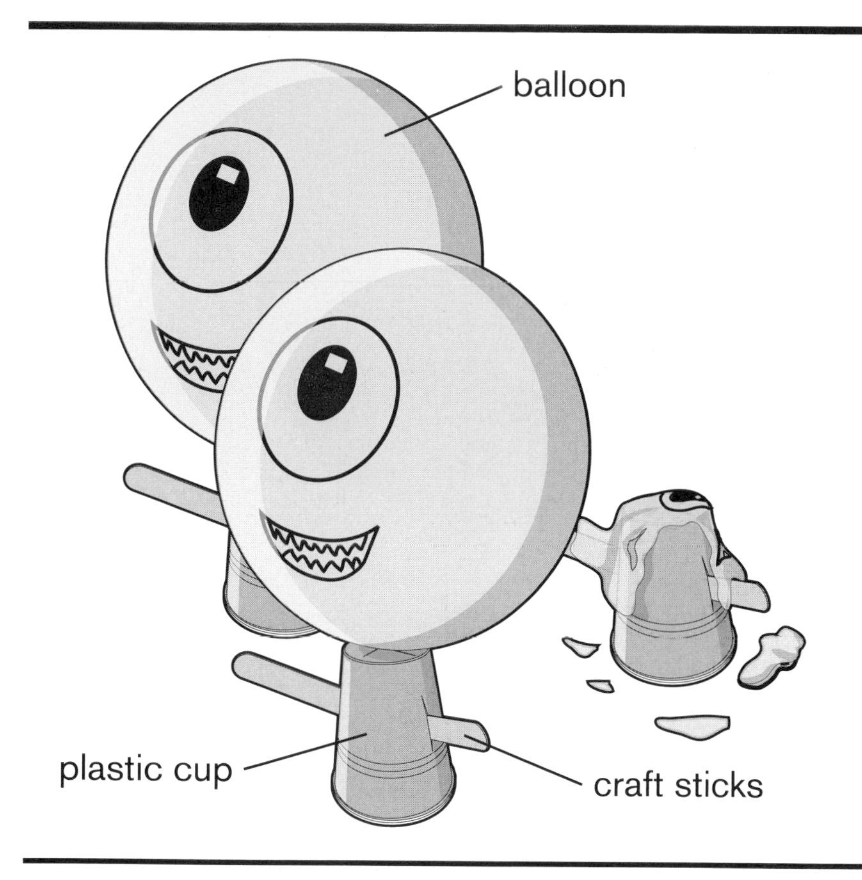


As H. G. Wells describes in *The War of the Worlds*, an alien invasion is inevitable, so we all must prepare with proper training. Bullets can't stop them, but they do have a weakness: Shoelace Blow Darts! These simple, pointy darts are capable of blowing the heads off an entire army of airheaded Martians.

### Supplies

2+ paper or plastic cup
1+ balloon
2+ craft sticks

### Tools

Hobby knife
Marker

## Step 1

Flip a plastic or paper cup upside down and cut two slits, forming an X, into the bottom center. Blow up your balloon and stick the tied end into the X on the bottom of the cup to hold the balloon in place. Next, cut two slits into the sides of the cup to insert the alien's craft-stick arms.

Don't forget to add the disturbing Mike Wazowski face with a marker!

# ANIMAL HUNT

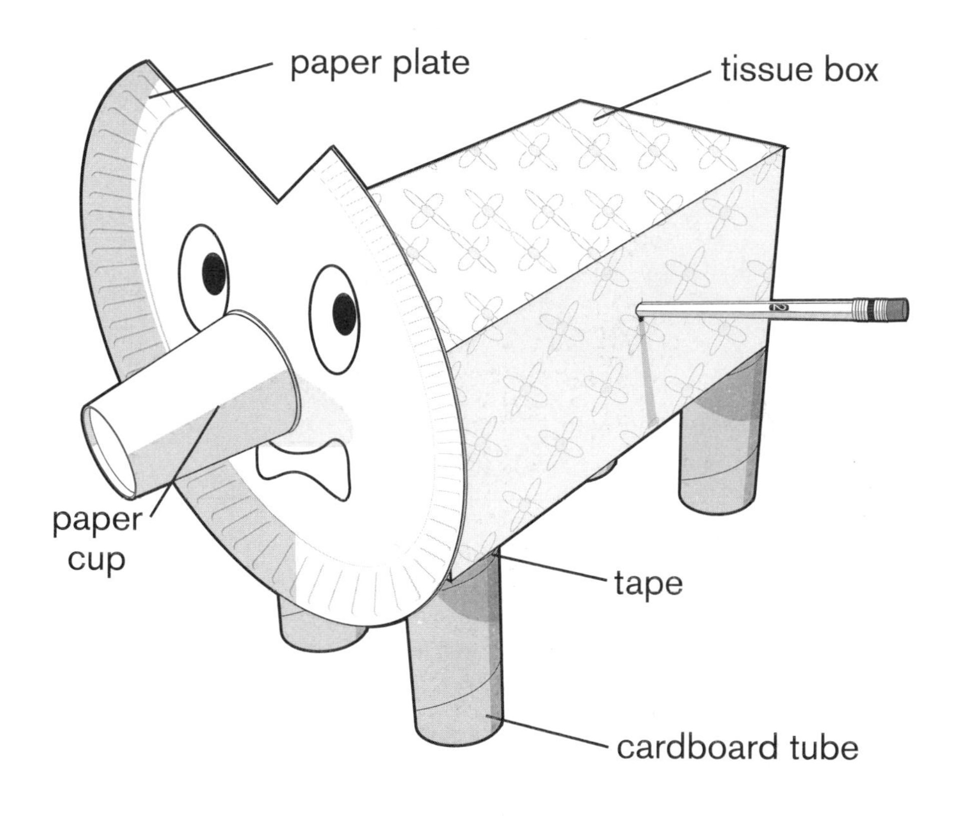


The legendary bloodsucking cardboard beast: *chupacabra*. Believed to exist only in the shadows of the unknown, this four-legged creature scavenges the countryside sniffing out its next prey with its ugly paper nose. You don't hunt it—it hunts *you*, as you carefully try to locate its private lair. You find its den entrance, and you enter. Are you ready?

### Supplies

1 tissue box
Tape (any kind)
4 toilet paper tubes
1 paper plate
1 paper cup

### Tools

Scissors
Marker

## Step 1

Flip over your tissue box. Tape the four cardboard tubes to the underside for the legs. Cut out a small wedge from your paper plate for the ear detail. Tape the paper-cup nose to the center, and then attach the head to the body using tape. Don't forget to add large fanged teeth and savage eyes.

# CEREAL TARGET

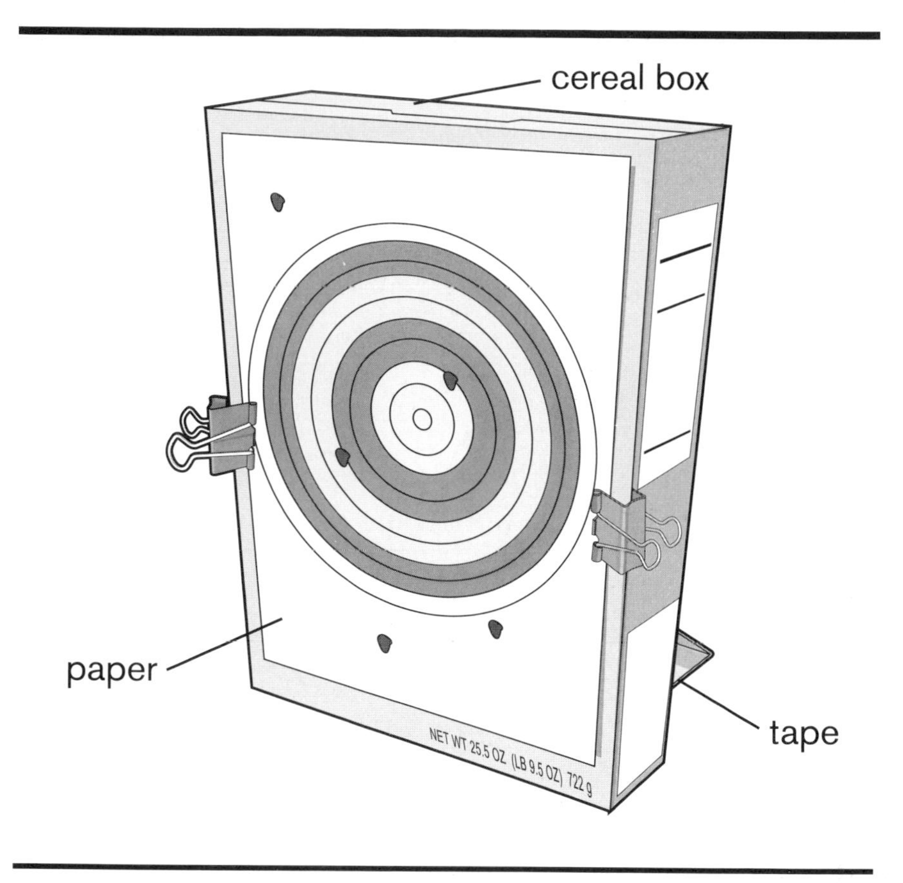


The Cereal Target is designed for the serious marksman perfecting his or her skills. A target lined with cardboard is perfect for stopping airsoft BBs or arrows made of writing utensils. With these interchangeable homemade targets, you'll soon master your miniweaponry.

### Supplies

1 cereal box
Tape (any kind)
2 binder clips (any size)
1 sheet of paper

### Tools

Hobby knife
Scissors

## Step 1

Remove the front panel of a cereal box with a hobby knife. Fold the removed section of cardboard in half and tape it to the back of the box for added support, as shown.

Cut two slits into the sides of the box for your binder clips. Clip custom-made targets to the box, or reproduce the targets on the following pages with a photocopier.

# OFFICIAL 10-FT TARGET

Weapon Choice ______________________________

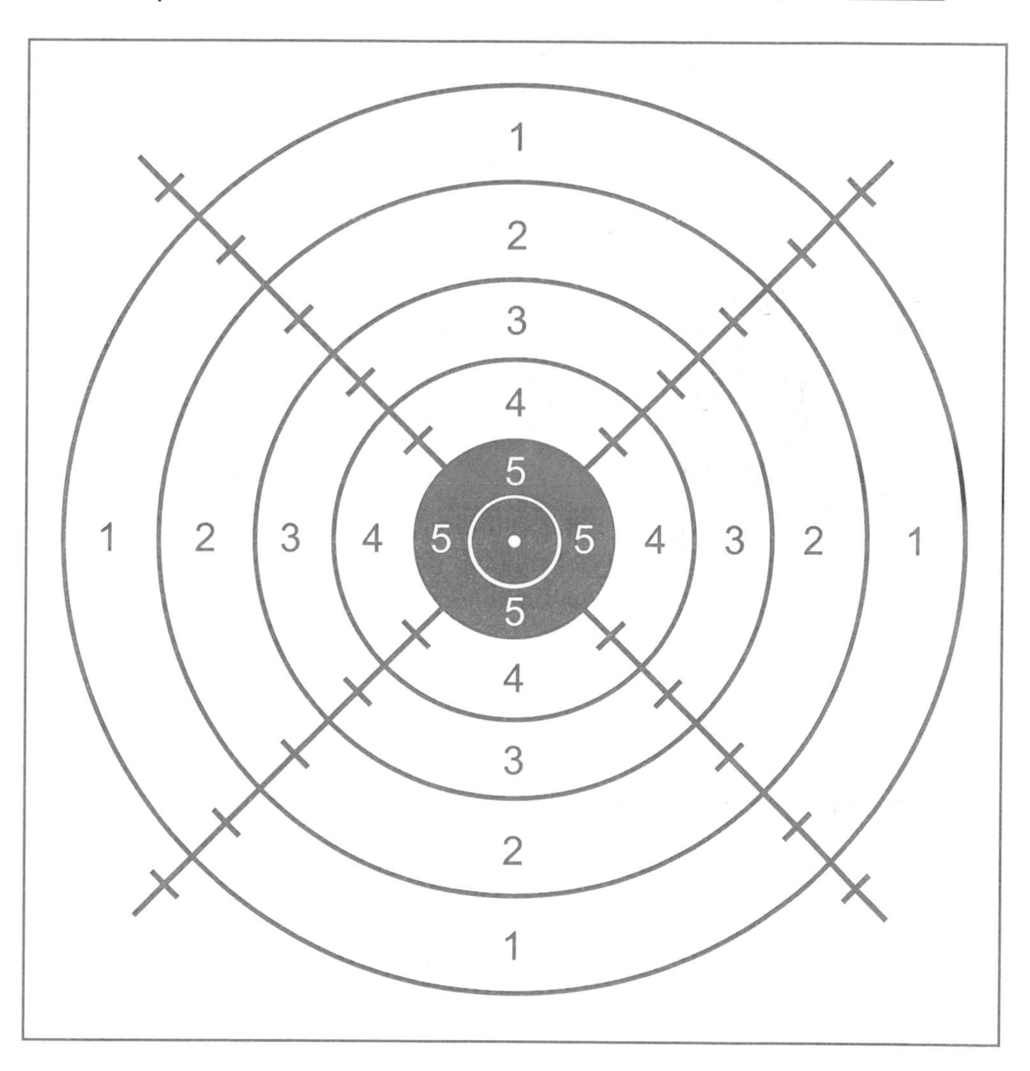


Competitor ______________________________ Date ______

Competitor Signature ______________________________

*Use a copy machine to make multiples and enlarge*

# DARTBOARD TARGET

Weapon Choice ______________________________

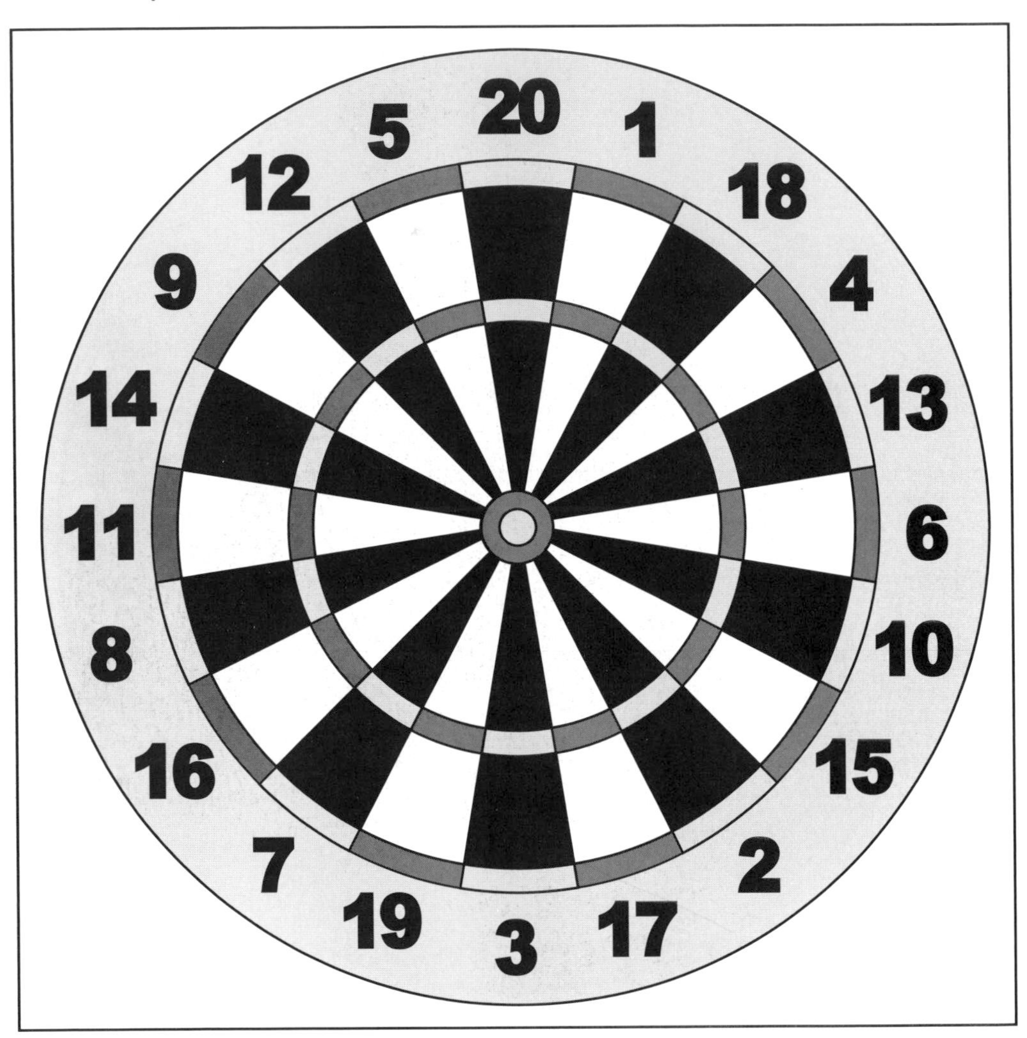


| | 20 | 19 | 18 | 17 | 16 | 15 | Bull |
|---|---|---|---|---|---|---|---|
| **P1** | ○○○ | ○○○ | ○○○ | ○○○ | ○○○ | ○○○ | ○○○ |
| **P2** | ○○○ | ○○○ | ○○○ | ○○○ | ○○○ | ○○○ | ○○○ |
| **P3** | ○○○ | ○○○ | ○○○ | ○○○ | ○○○ | ○○○ | ○○○ |

Competitor ______________________________ Date ______

Competitor Signature ______________________________



*Use a copy machine to make multiples and enlarge*

# ZOMBIE TARGET

*Use a copy machine to make multiples and enlarge*

**For more information and free downloadable targets, please visit:**

**MINIWEAPONSBOOK.COM**

***DON'T FORGET TO JOIN THE MINIWEAPONS ARMY ON FACEBOOK:***

**Miniweapons of Mass Destruction: Homemade Weapons Page**

# ALSO FROM CHICAGO REVIEW PRESS

## Miniweapons of Mass Destruction 2

**Build a Secret Agent Arsenal**

John Austin

978-1-56976-716-0
$16.95 (CAN $18.95)

As a budding spy, what better way to conceal your clandestine activities than to miniaturize your secret agent arsenal? Hide a mini catapult in a breath mint tin. Turn a Tic Tac case into vest-pocket candy shooter. Or transform a milk jug cap into a fake wristwatch that launches tiny paper darts. Toy designer and author John Austin provides detailed, step-by-step instructions with diagrams to show James Bondiacs how to build 35 different spy weapons and surveillance tools, including:

- Crayon Catapult
- Grappling Gun
- Detonating Pen
- Vest Pocket Mini
- Cotton Swab .38 Special
- Gift Card Coin Launcher
- Candy Glock 33
- Double-Barreled Band Gun

Once you have assembled your arsenal, you'll learn how to hide it–inside a deck of cards, a false-bottom soda bottle, or a cereal box briefcase. *Miniweapons of Mass Destruction 2* also includes targets for practicing your spycraft, including a flip-down firing range, a fake security camera, and sharks with laser beams. And if you think yourself more of an evil genius in training, this book also has projects to keep you busy while you finish planning your volcano lair–a Q-pick Blowgun, a Paper Throwing Star, a Bowler Hat Launcher, and more. Fluffy Persian cat not included.

## So Now You're a Zombie

**A Handbook for the Newly Undead**

John Austin

978-1-56976-342-1
$14.95 (CAN $16.95)

Zombies know that being undead can be disorienting. Your arms and other appendages tend to rot and fall off. It's difficult to communicate with a vocabulary limited to moans and gurgles. And that smell! (Yes, it's *you*.) But most of all, you must constantly find and ingest human brains. *Braaaains!!!*

What's a reanimated corpse to do?

As the first handbook written specifically for the undead, *So Now You're a Zombie* explains how your new, putrid body works and what you need to survive in this zombiphobic world. Dozens of helpful diagrams outline attack strategies to secure your human prey, such as the Ghoul Reach, the Flanking Zeds, the Bite Hold, and the Aerial Fall. You'll learn how to successfully extract the living from boarded-up farmhouses and broken-down vehicles. Zombiologist John Austin even explores the upside of being a zombie. Gone are the burdens of employment, taxes, social networks, and basic hygiene, allowing you to focus on the simple necessities: the juicy gray matter found in the skulls of the living.

## Backyard Ballistics

**Build Potato Cannons, Paper Match Rockets, Cincinnati Fire Kites, Tennis Ball Mortars, and More Dynamite Devices**

William Gurstelle

978-1-55652-375-5
$16.95 (CAN $18.95)

"If you want to make a potato souffle, pick up a book by Julia Child. If you want to decorate your holiday cards with hand-cut potato stamps, look to a Martha Stewart manual. If, however, you'd like to launch a potato in a blazing fireball of combusting hairspray from a PVC pipe, your best source is *Backyard Ballistics*." —*Time Out New York*

Ordinary folks can construct 13 awesome ballistic devices in their garage or basement workshops using inexpensive household or hardware-store materials and this step-by-step guide. Clear instructions, diagrams, and photographs show how to build projects ranging from the simple—a match-powered rocket—to the more complex—a scale-model tabletop catapult—to the offbeat—a tennis-ball cannon.

## The Art of the Catapult

**Build Greek Ballistae, Roman Onagers, English Trebuchets, and More Ancient Artillery**

William Gurstelle

978-1-55652-526-1
$16.95 (CAN $18.95)

"This book is a hoot . . . the modern version of *Fun for Boys* and *Harper's Electricity for Boys*." —*Natural History*

Whether playing at defending their own castle or simply chucking pumpkins over a fence, wannabe marauders and tinkerers will become fast acquainted with Ludgar, the War Wolf, Ill Neighbor, Cabulus, and the Wild Donkey—ancient artillery devices known commonly as catapults. Instructions and diagrams illustrate how to build seven authentic, working model catapults, including an early Greek ballista, a Roman onager, and the apex of catapult technology, the English trebuchet.

## Soda-Pop Rockets

**20 Sensational Rockets to Make from Plastic Bottles**

Paul Jarvis

978-1-55652-960-3
$16.95 (CAN $18.95)

Anyone can recycle a plastic bottle by tossing it into a bin, but it takes a bit of skill to propel it into a bin from 500 feet away. This fun guide features 20 different easy-to-launch rockets that can be built from discarded plastic drink bottles. After learning how to construct and launch a basic model, you'll find new ways to modify and improve your designs. Clear, step-by-step instructions with full-color illustrations accompany each project, along with photographs of the author firing his creations into the sky.

## The Way Toys Work

**The Science Behind the Magic 8 Ball, Etch A Sketch, Boomerang, and More**

Ed Sobey and Woody Sobey

978-1-55652-745-6
$14.95 (CAN $16.95)

"Perfect for collectors, for anyone daring enough to build homemade versions of these classic toys and even for casual browsers." —*Booklist*

Profiling 50 of the world's most popular playthings—including their history, trivia, and the technology involved—this guide uncovers the hidden science of toys. Discover how an Etch A Sketch writes on its gray screen, why a boomerang returns after it is thrown, and how an RC car responds to a remote control device.

**Available at your favorite bookstore, by calling (800) 888-4741, or at www.chicagoreviewpress.com**